高等学校教材

高分子专业英语
Professional English of Polymer

张海磊　主　编
张元功　房丽萍　副主编

化学工业出版社
·北京·

内容简介

《高分子专业英语》分为基础篇和拓展篇。基础篇中，精选25个中心词，以中心词为基础进而拓展到其派生词、同（近）义词、相关词和词组，以加深读者对高分子词汇的认识和理解；并根据相应内容，在部分章节中加入了拓展阅读、写作知识、背景介绍、课后问答题等，融入高分子科学的发展、高分子领域的诺贝尔奖成果和重要研究进展、趣味故事、学术规范等内容，开阔读者视野。拓展篇中，根据升学、工作、投稿、参会、申请等具体需要，总结了英文自我介绍、英文信件写作、参会常用英文表达及英文期刊投稿等相关内容，以提高实用性。

本书提供4套模拟测试题及答案，供测评或自测使用。附录部分总结了中国科学院高分子科学（Polymer Science）小类分项下期刊的主页和投稿地址，便于读者使用。生词表中除了汇总本书各小节中的中心词及其相关词语外，还汇总了例句和短文中的相关专业词汇。

本书可作为高等学校高分子材料与工程、高分子化学与物理以及材料化学、化学等专业的教材，也可供从事高分子科学研究、开发和应用的人员参考。

图书在版编目（CIP）数据

高分子专业英语/张海磊主编；张元功，房丽萍副主编. —北京：化学工业出版社，2024.4
ISBN 978-7-122-45129-3

Ⅰ. ①高… Ⅱ. ①张… ②张… ③房… Ⅲ. ①高分子材料-英语-高等学校-教材 Ⅳ. ①TB324

中国国家版本馆CIP数据核字（2024）第043660号

责任编辑：提　岩　　　　　　　文字编辑：曹　敏
责任校对：李雨晴　　　　　　　装帧设计：王晓宇

出版发行：化学工业出版社
　　　　（北京市东城区青年湖南街13号　邮政编码100011）
印　　装：河北鑫兆源印刷有限公司
787mm×1092mm　1/16　印张9¾　字数235千字
2024年8月北京第1版第1次印刷

购书咨询：010-64518888　　　　售后服务：010-64518899
网　　址：http://www.cip.com.cn
凡购买本书，如有缺损质量问题，本社销售中心负责调换。

定　　价：30.00元　　　　　　　　　版权所有　违者必究

本书编审人员

主　编：张海磊
副主编：张元功　房丽萍
参　编：巴信武　武永刚　白利斌　商　闯
　　　　宋伟华　高红燕
主　审：闫明涛

前言

高分子科学是一门内容广泛，且与多学科交叉渗透、相互关联的综合性学科。进入 21 世纪，我国在高分子相关科研、产业、教育等领域的国际合作日益增多。高分子科学在新时代的发展不仅需要具备高分子化学、高分子物理、高分子材料成型加工等方面基础知识的专业人才，更需要兼具扎实专业外语水平和国际化视野的高水平人才。因此，《高分子专业英语》教材既要注重基础词汇的积累，又要体现实用性，使之既可以作为教材，又可以作为师生或企事业单位相关工作人员参加国际学术会议、开展国际合作、投稿英文期刊、申请海外院校以及日常使用的工具书。

河北大学是教育部与河北省人民政府"部省合建"的综合性院校。2013 年化学学科获批为河北省国家重点学科培育项目，2016 年列入河北省世界一流学科建设项目。高分子材料化学教研室隶属于河北大学化学与环境科学学院，承担着材料化学、高分子材料与工程专业本科生和高分子化学与物理专业研究生的教学工作，同时服务于河北省重点学科高分子化学与物理的建设。本教材主编张海磊副教授，为 Ghent University 访问学者，主持国家自然科学基金、国家留学基金公派访问学者项目、中央引导地方科技发展资金项目、河北省自然科学基金、河北省课程思政示范项目、河北省高等学校科学技术研究项目、河北省研究生示范课程立项建设项目等多项课题，近年来以第一/通讯作者身份在 Nature Communications 等期刊发表 SCI 检索论文 30 余篇，为省级教学团队负责人。在此背景下，作者结合高分子科学的发展现状，系统整理了高分子专业外语相关知识点，组织编写本教材。本教材在加深学生对高分子科学基础知识理解的同时，增加了实用性、趣味性和前沿性的内容，以期提高学生学习高分子专业英语的兴趣，拓展学生的视野。

本书由张海磊担任主编，张元功、房丽萍担任副主编。Unit 1～Unit 13、Unit 21～Unit 25、Unit 28～Unit 30 和模拟测试由张海磊编写；Unit 14～Unit 17 由张元功编写；Unit 18～Unit 20 由宋伟华、高红燕编写；Unit 26 和 Unit 27 由张海磊、商闯编写；附录由张元功、房丽萍编写。全书由张海磊、张元功、房丽萍统稿，闰明涛教授主审，巴信武、武永刚、白利斌就统稿工作中的若干学术问题和内容的取舍提出了宝贵意见。本书在编写过程中参考了国内优秀的高分子专业英语相关教材，在此对相关作者深表感谢。本书的出版得到了河北省课程思政示范项目、中央引导地方科技发展资金项目、河北省科技特派员项目和河北大学的大力支持，在此也一并致谢。

由于编者学识水平所限，书中不足之处在所难免，敬请广大读者批评指正！

编者
2024 年 1 月

目录

基 础 篇

Unit 1	Polymer	002
Unit 2	Molecular Weight	005
Unit 3	Synthesis	008
Unit 4	Radical	010
Unit 5	Monomer	013
Unit 6	Solvent	016
Unit 7	Purification	019
Unit 8	Reaction	022
Unit 9	Characterization	025
Unit 10	Processing	027
Unit 11	Property	030
Unit 12	Degradation	033
Unit 13	Modulus	036
Unit 14	Morphology	039
Unit 15	Element	042
Unit 16	Function	044
Unit 17	Instrument	046
Unit 18	Natural	048
Unit 19	Product	051
Unit 20	Data	054
Unit 21	Symbol	057
Unit 22	Abbreviation	060
Unit 23	Configuration	063
Unit 24	Conformation	066
Unit 25	Crystal	068

拓 展 篇

Unit 26	Introduce Yourself	072
Unit 27	Write a Letter	078
Unit 28	Submission	085
Unit 29	Academic Meeting	090
Unit 30	Safety	097

模拟测试

模拟测试题（一） ··· 106
模拟测试题（二） ··· 108
模拟测试题（三） ··· 110
模拟测试题（四） ··· 112
模拟测试题（一）答案及评分标准 ··· 114
模拟测试题（二）答案及评分标准 ··· 116
模拟测试题（三）答案及评分标准 ··· 118
模拟测试题（四）答案及评分标准 ··· 120

附　录

高分子领域常见外文 SCI 期刊信息汇总 ·· 123
生词表（Vocabulary） ·· 126

参考文献 ·· 146

基础篇

Unit 1
Polymer

- 读音：英 /ˈpɒlɪmə(r)/　美 /ˈpɑːlɪmər/
- 释义：*n.* 聚合物
- 英英释义：A chemical substance consisting of large molecules made from many smaller and simpler molecules (Cambridge Dictionary).

例　句

- The novel polymer relies on carbon-sulfur bonds within the material.
  新的聚合物是以材料中的碳硫键为基础。
- In many cases polymer chains are linear.
  在许多情况下聚合物链是线型的。
- As the solvent evaporated, they were left with model compound mixed into a polymer film.
  等到溶剂挥发后，就得到混合有模型化合物的聚合物薄膜。

背景知识

　　Polymer（Fig.1-1）由 poly 和 mer 两个单元构成，poly 和 mer 的来源为希腊文，其中 poly 指的是多、复、聚的意思，mer 代表的意思是部分，poly + mer 字面意思为多个部分。

同（近）义词：

| macromolecule | 英 /ˌmækrəʊˈmɒləkjuːl/　美 /ˈmækroʊˌmɑːlɪkjuːl/ | *n.* | 大分子 |

派生词及相关词汇：

homopolymer	英 /ˌhɒməʊˈpɒlɪmə(r)/　美 /hoʊməˈpɑːləmər/	*n.*	均聚物
copolymer	英 /kəʊˈpɒlɪmə(r)/　美 /koʊˈpɑːləmər/	*n.*	共聚物
dipolymer	英 /ˈdaɪpəliːmə(r)/　美 /daɪˈpɑːlɪmər/	*n.*	二元共聚物
terpolymer	英 /tɜːˈpɒlɪmə/　美 /tərˈpɑːlɪmər/	*n.*	三元共聚物
polymerization	英 /ˌpɒlɪməraɪˈzeɪʃn/　美 /ˌpɑːlɪmərəˈzeɪʃn/	*n.*	聚合
polymeric	英 /ˌpɒlɪˈmerɪk/　美 /ˌpɑːlɪˈmerɪk/	*adj.*	聚合的
macromolecular	英 /ˌmækrəʊməˈlekjulə/　美 /ˌmækroʊməˈlɪkjuːl/	*adj.*	大分子的
dendrimer	英 /ˈdendrəmə(r)/　美 /ˈdendrɪmər/	*n.*	树枝状聚合物

固定搭配/常用短语:

dead polymer	无活性聚合物
linear polymer	线型聚合物
branched polymer	支化聚合物
dendritic polymer	树枝状聚合物
chain polymerization	链式聚合
step-growth polymerization	逐步聚合
ionic polymerization	离子聚合
bulk polymerization	本体聚合
interfacial polymerization	界面聚合
emulsion polymerization	乳液聚合

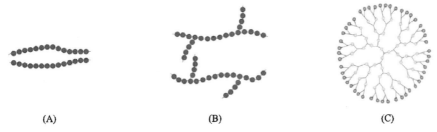

Figure1-1　Linear polymer (A), branched polymer (B) and dendritic polymer (C)

Question (Please answer in English):

What are the differences and similarities between branched polymer and dendritic polymer?

拓展阅读

Hermann Staudinger and the Origin of Macromolecular Chemistry

Hermann Staudinger is a famous chemist and is praised as the "father of macromolecular chemistry". He achieved the Nobel Prize for Chemistry on Dec 10th, based on his pioneering research. The headline "High Polymers bring High Honors" became famous worldwide. Staudinger discovered the molecular blueprints of synthetic and natural polymeric materials with high molecular weight. His concept is very revolutionary. In his concept, the macromolecule is constituted by a large number of small monomer molecules linking by covalent bonds. This point marked the beginning of the new era of molecular design of high molecular weight structural and functional polymeric materials. At the beginning of the 1900s, some scholars believed that the measured high molecular weights were only apparent values caused by the aggregation of small molecules into colloids. They refused to accept the possibility that small molecules could link together covalently to form high molecular weight compounds. Most early imitations exhibited rather poor product qualities as compared to those of the corresponding natural products. It was Staudinger's molecular blueprint that was applied successfully to design novel polymers with property unparalleled in nature. Further evidence to support his concept emerged in the 1930s. High molecular weights of polymers were confirmed by membrane osmometry, and also by Staudinger's measurements of viscosity in solution. His pioneering research has afforded the world myriad plastics, textiles, and other polymeric materials. Nowadays, high molecular weight polymer products are integral parts of our daily life and secure high quality of life. Polymeric materials are superior with respect to their unique combination of flexible feedstocks supply , attractive cost/performance ratio, desirable ecoefficiency, low energy demand during preparation and processing, ease of processing with short cycle times suitable for the modern industrial mass production, extraordinary property versatility, wide application range, and recoverability. No other class of materials is able to match similarly diversified property. Staudinger's concept will continue to support the solid and still expanding base on which we build new structural and functional polymers.

Unit 2
Molecular Weight

- 读音：molecular 英 /məˈlekjələ(r)/　美 /məˈlekjələr/
  weight 英/weɪt/　美/weɪt/
  molecular　*adj*. 分子的　weight　*n*. 重量
- 释义：molecular weight　分子量
- 英英释义：The total of the relative atomic masses of the atoms in a particular molecule (Cambridge Dictionary).

例 句

- This is because higher molecular weight molecules are less soluble in water.
  这是由于更高分子量的分子在水中溶解更少。
- Enzymes are large molecular weight compounds.
  酶是高分子化合物。
- Polyacrylate sodium with high molecular weight is commonly-used as a library of important anionic flocculation in industry.
  高分子量聚丙烯酸钠是一类常用的阴离子型絮凝剂，在工业上有着广泛的应用。

同（近）义词组：

relative molecular mass
n. 相对分子质量（分子量）
*提示: relative molecular mass 一般用于描述具有准确分子量的小分子，与 polymer 或者 macromolecule 搭配使用的情况较为少见。

固定搭配/常用短语：

number-average molecular weight	数均分子量
weight-average molecular weight	重均分子量
viscosity-average molecular weight	黏均分子量
repeating unit	重复单元
degree of polymerization	聚合度
hydrodynamic radius	流体动力学半径

派生词及相关词汇：

polydispersity	英 /ˌpɒlɪdɪs'pɜːsɪtɪ/ 美 /ˌpɔlidis'pəːsiti/	n.	多分散性
chromatography	英 /ˌkrəʊmə'tɒgrəfi/ 美 /ˌkroʊmə'tɑːgrəfi/	n.	色谱法
gel	英 /dʒel/ 美 /dʒel/	n.	凝胶
permeation	英 /ˌpɜːmi'eɪʃn/ 美 /ˌpɜːrmi'eɪʃn/	n.	渗透
penetration	英 /ˌpenə'treɪʃ(ə)n/ 美 /ˌpenə'treɪʃ(ə)n/	n.	穿透
exclusion	英 /ɪk'skluːʒ(ə)n/ 美 /ɪk'skluːʒn/	n.	排除

例 句

- Gel permeation chromatography (GPC) was used to determine the molecular weight of the polymer.
  利用凝胶渗透色谱(GPC)测定了聚合物的分子量。
- Gel permeation chromatography (GPC) is also known as size-exclusion chromatography (SEC).
  凝胶渗透色谱(GPC)又被称作体积排除色谱(SEC)。
- Polydispersity index (PDI) is calculated based on the ratio of number-average molecular weight (M_n) to weight-average molecular weight (M_w).
  多分散性指数是根据数均分子量和重均分子量的比值求得的。

拓展阅读

Gel Permeation Chromatography

Gel permeation chromatography (GPC), also known as size-exclusion chromatography (SEC) is a commonly used instrument that separates analytes on the basis of size. The technique is usually employed in analysis of polymers and allows for the determination of number-average molecular weight (M_n), weight-average molecular weight (M_w), polydispersity and other related parameters. This technique is based on the penetration of molecules into the cavities of macroporous support, mostly made from hydrophilic gels. In general, molecules with a hydrodynamic diameter smaller than the diameter of the pores in the support can enter the pores more easily and therefore spend more time in these pores, whereas molecules with larger diameters are excluded, thus eluting more quickly. Therefore, large molecules are excluded from the pores of the gels and are eluted first. If an analyte is too large, it will not be retained; conversely, if the analyte is too small, it may be retained completely. This technique can easily be automated and has allowed for the quick estimation of molecular weight, as well as distribution for polymer samples. As can be inferred, there is a limited range of molecular weight that can be separated by each column, and therefore the size of the pore for the packing should be chosen according to the range of molecular weight of analyte to be separated (Fig.1-2).

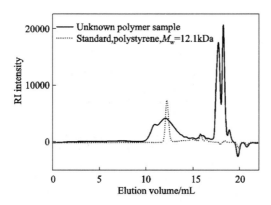

Figure1-2　Elution curves measured from GPC

Questions (Please answer in English):
(1) What is the relationship between degree of polymerization and molecular weight?
(2) How to calculate the value of polydispersity index?

Unit 3
Synthesis

- 读音：英 /ˈsɪnθəsɪs/　美 /ˈsɪnθəsɪs/
- 释义：*n.* 合成
- 英英释义：The production of a substance from simpler materials after a chemical reaction (Cambridge Dictionary).

例 句

- The synthesis of polymethyl methacrylate was reported in this paper.
  本文报道了聚甲基丙烯酸甲酯的合成。
- The synthesis of the p-conjugated polymer catalyzed by Pd(0) was investigated.
  研究了零价钯催化合成共轭聚合物的合成工艺。
- The synthesis method of chitosan hydrogel and its application in medicine, anti-bacteria and adsorption has been discussed.
  讨论了壳聚糖水凝胶的合成方法及其在医药、抗菌、吸附等方面的应用。

派生词及相关词汇：

synthesize	英 /ˈsɪnθəsaɪz/　美 /ˈsɪnθəsaɪz/	*v.*	合成
synthesized	英 /ˈsɪnθəsaɪzd/　美 /ˈsɪnθəsaɪzd/	*adj.*	合成的
synthetic	英 /sɪnˈθetɪk/　美 /sɪnˈθetɪk/	*adj.*	合成的，人造的
		n.	合成物
manmade	英 /ˈmænˈmeɪd/　美 /ˈmænˈmeɪd/	*adj.*	人造的
		n.	人工制品
natural	英 /ˈnætʃ(ə)rəl/　美 /ˈnætʃərəl/	*adj.*	天然的

* 提示：synthetic (*adj.* 合成的、人造的) 与 natural (*adj.* 天然的) 互为反义词，与 polymer 联用时，分别指代合成高分子和天然高分子。manmade 也可用于与 polymer 联用，代指人造高分子或合成高分子，但常见于科普性文章或口语表达。

固定搭配/常用短语：

chemical synthesis	化学合成
hydrothermal synthesis	水热合成
one-pot synthesis	一锅法合成

Unit 3　Synthesis

（续表）

solid-phase synthesis	固相合成
synthetic rubber	合成橡胶
synthetic fiber	人造纤维
synthetic resin	合成树脂
synthetic material	合成材料
synthetic process	合成过程
synthetic reaction	合成反应

同（近）义词：

preparation	英 /ˌprepəˈreɪʃn/　美 /ˌprepəˈreɪʃn/	n.	制备

*提示: synthesis 和 preparation 均是高分子专业相关文献中的高频词，也是论文题目和图例中经常用到的词汇，经常放在论文题目的句首使用，表示某种物质的合成或制备。一般而言，在描述涉及化学反应类的聚合物的生成过程时，使用 synthesis；用于描述聚合物共混、改性、加工成型等方面研究时，使用 preparation 较为合适。

例如：

- **Synthesis** of Poly(Methyl Methacrylate)-based Polyrotaxane via Reversible Addition-Fragmentation Chain Transfer Polymerization (Wang Y, *et al.*, *ACS Macro Letters*, 2020, 9: 1853-1857).
  基于可逆加成-断裂链转移聚合合成聚甲基丙烯酸甲酯基聚轮烷
- **Preparation** of Highly Alkaline Stable Poly(Arylene-Imidazolium) Anion Exchange Membranes Through an Ionized Monomer Strategy (Xue B, *et al.*, *Macromolecules*, 2021, 54: 2202-2212).
  通过电离单体策略制备高碱稳定性聚（亚芳基-咪唑）阴离子交换膜

拓展阅读

A Typical Synthesis Procedure

A mixture of monomer-1 (2.0g, 3.1mmol), monomer-2 (0.1g, 0.3mmol), $NaHCO_3$ (0.4g, 4.8mmol), H_2O (4mL), and tetrahydrofuran (20mL) was carefully degassed before $Pd(PPh_3)_4$ (8mg, 0.082mmol) was added. The mixture was refluxed for 1 day under stirring. Water and CH_2Cl_2 (50mL) were added, the organic layer separated, the aqueous layer extracted with CH_2Cl_2, and the combined organic layers dried over Na_2SO_4. Removal of the solvent, the crude product was chromatographically purified on silica gel eluting with CH_2Cl_2/hexane (1:1, v/v) increasing to CH_2Cl_2 to afford the polymeric product as a white solid (0.11g, 30%).

(Wu Y, *et al.*, *Macromolecules*, 2010, 43, 731-738)

Unit 4
Radical

- 读音：英 /ˈrædɪkl/ 美 /ˈrædɪkl/
- 释义：*n*. 自由基
- 英英释义：A molecule that has an extra electron and therefore reacts very easily with other molecules (Cambridge Dictionary).

例 句

- When a free radical reacts with vinyl-based monomer a new radical is created, and consequently chain reactions are often initiated.
  当一个自由基与乙烯基单体反应时，会产生一个新的自由基，因此往往会引发链式反应。
- Hydroquinone is usually used as radical scavenger.
  对二苯酚常被用作自由基清除剂。
- A stable radical intermediate is a necessary requisite to loss of carbon dioxide.
  一个稳定的自由基中间体是脱除二氧化碳所需的必要条件。

固定搭配/常用短语：

free radical	自由基
free radical polymerization	自由基聚合反应
living radical polymerization	活性自由基聚合
radical reaction	自由基反应
radical scavenger	自由基清除剂
carbon (free) radical	碳自由基
oxygen (free) radical	氧自由基

*提示：注意 oxygen radical 和 reactive oxygen species (ROS) 是两个概念，oxygen radical 指的是氧自由基，ROS 指的是活性氧。氧自由基是活性氧中的一类，但活性氧还包括很多其他物质，比如过氧化物、超氧离子等。

派生词及相关词汇：

自由基聚合反应中包含链引发（initiation）、链增长（propagation）、链终止（termination）的过程。链引发的方式分很多种，包括光引发、热引发、引发剂受热分解等；链终止又分为偶合终止和歧化终止，这些都是与自由基具有很密切关系的词汇。

Unit 4 Radical

initiation	英 /ɪˌnɪʃi'eɪʃn/ 美 /ɪˌnɪʃi'eɪʃn/	n.	（链）引发
initiate	英 /ɪ'nɪʃieɪt/ 美 /ɪ'nɪʃieɪt/	v.	引发
initiator	英 /ɪ'nɪʃieɪtə(r)/ 美 /ɪ'nɪʃieɪtər/	n.	引发剂
homolysis	英 /hɒ'mɒlɪsɪs/ 美 /hɒ'mɒlɪsɪs/	n.	均裂
photoinitiator	英 /fəʊtwɑ:'nɪʃɪeɪtə/ 美 /fəʊtwɑ:'nɪʃɪeɪtə/	n.	光引发剂
photoinitiation	英 /fəʊtwɑ:nɪʃ'ɪeɪʃn/ 美 /fəʊtwɑ:nɪʃ'ɪeɪʃn/	n.	光引发
inimer	英 /ɪnɪmə/ 美 /ɪnɪmə/	n.	引发剂型单体
propagation	英 /ˌprɒpə'geɪʃn/ 美 /ˌprɑ:pə'geɪʃn/	n.	（链）增长
propagate	英 /'prɒpəgeɪt/ 美 /'prɑ:pəgeɪt/	v.	增长
termination	英 /ˌtɜːmɪ'neɪʃn/ 美 /ˌtɜːrmɪ'neɪʃn/	n.	（链）终止
terminate	英 /'tɜːmɪneɪt/ 美 /'tɜːrmɪneɪt/	v.	终止
combination	英 /ˌkɒmbɪ'neɪʃn/ 美 /ˌkɑ:mbɪ'neɪʃn/	n.	偶合
disproportionation	英 /ˌdɪsprəˌpɔ:ʃə'neɪʃ(ə)n/ 美 /dɪsprəpɔ:ʃə'neɪʃən/	n.	歧化反应
inhibitor	英 /ɪn'hɪbɪtə(r)/ 美 /ɪn'hɪbɪtər/	n.	阻聚剂
redox	英 /'ri:dɒks; 'redɒks/ 美 /'ri:ˌdɑ:ks/	n.	氧化还原
species	英 /'spi:ʃi:z/ 美 /'spi:ʃi:z/	n.	种类

常见的自由基聚合引发剂：

- azobisisobutyronitrile *n.* 偶氮二异丁腈；abbr. AIBN
  英 /əzɒbaɪsɪsəʊ'bʌtɪrəni:traɪ/ 美 /əzɑ:baɪsɪsoʊ'bʌtɪrənitraɪ/
- benzoyl peroxide 过氧化苯甲酰；abbr. BPO
  benzoyl 英 /'benzəʊɪl/ 美 /'benzoʊˌɪl/ *n.* 苯甲酰
  peroxide 英 /pə'rɒksaɪd/ 美 /pə'rɑ:ksaɪd/ *n.* 过氧化氢，过氧化物
- Fenton reagent 芬顿试剂
- persulphate *n.* 过硫酸盐
  英 /pə'sʌlfeit/ 美 /pə'sʌlfeit/

拓展阅读

Redox Polymerization

A separate initiation step is essentially required in all free-radical polymerizations in which a radical species is generated in the reaction mixture. Some types of chain polymerizations are initiated by adding a stable free radical, also known as radical initiator, such as azobisisobutyronitrile (AIBN) and benzoyl peroxide (BPO). Generally, it shows little or no tendency for self-combination, directly to the reactants, but a separate initiation step is still involved. Radical initiation reactions, therefore, can be divided into two general types according to the manner in which the first radical species is formed: (i) homolytic decomposition of covalent bonds by energy absorption; (ii) electron transfer from ion or atom containing unpaired electron followed by bond dissociation in the acceptor molecule. For the homolysis of covalent bonds of most practical thermal initiators, the bond dissociation energy required is in a certain range, and compounds with energy below or above this range give either too fast or too slow a rate of generation of radicals at the polymerization temperatures generally used. A very effective method of generating free radicals under mild conditions is by one electron transfer reactions, the most effective of which is redox initiation. Redox initiation occurs when a reducing agent (or system) is mixed with an oxidizing agent (or system) generating reactive species that allow the polymerization of the surrounding monomers. This method has found broad applications for initiating polymerization reactions and has industrial importance, especially in low temperature emulsion polymerization systems and grafting techniques.

Unit 5
Monomer

- 读音：英 /ˈmɒnəmə(r)/ 美 /ˈmɑːnəmər/
- 释义：*n.* 单体
- 英英释义：A compound whose molecules can join together to form a polymer（《柯林斯英汉双解大词典》）.

例 句

- Some methods for removal of residual monomers from polymer products were listed as follows.
  从聚合物产品中去除残留单体的方法如下。
- The essential feature of monomer molecules is the ability to form chemical bonds with at least two other monomer molecules.
  单体分子的基本特征是能够与至少两个其他单体分子形成化学键。
- The monomer for the preparation of epoxy resin is a light amber fluid which is usually quite viscous.
  用于制备环氧树脂的单体是一种淡琥珀色的液体，通常很黏。

派生词及相关词汇：

monomeric	英 /ˌmɒnəˈmerɪk/ 美 /ˌmɑːnəˈmerɪk/	*adj.*	单体的
unsaturated	英 /ʌnˈsætʃəˌreɪtɪd/ 美 /ʌnˈsætʃəˌreɪtɪd/	*adj.*	不饱和的
olefin	英 /ˈəʊləfɪn/ 美 /ˈoʊləfɪn/	*n.*	烯烃
olefinic	英 /ˌəʊləˈfɪnɪk/ 美 /ˌoʊləˈfɪnɪk/	*adj.*	不饱和的
vinyl	英 /ˈvaɪnl/ 美 /ˈvaɪnl/	*n.*	乙烯基
bifunctional	英 /baɪˈfʌŋkʃən(ə)l/ 美 /bɪˈfʌŋkʃənl/	*adj.*	双官能团的
monofunctional	英 /ˌmɒnəʊˈfʌŋkʃənəl/ 美 /ˌmɔːnəʊˈfʌŋkʃənəl/	*adj.*	单官能团的
multifunctional	英 /ˌmʌltiˈfʌŋkʃənl/ 美 /ˌmʌltiˈfʌŋkʃənl/	*adj.*	多官能团的

固定搭配/常用短语：

vinyl-based monomer	乙烯基单体
bifunctional monomer	双官能[基]单体
A_xB_y-type monomer	A_xB_y 型单体

例 句

- Generally, an inimer can play dual roles in polymerizations, which can serve as a monomer and an initiator (Liu F, *et al.*, *Polymer Chemistry*, 2018, 9: 5024-5031).
  通常，在聚合过程中，引发剂型单体可以起到双重作用，既可以作为单体，也可以作为引发剂。
- Multifunctional monomers may serve as a good candidate for synthesis of hyperbranched polymers.
  多功能单体可作为合成超支化聚合物的良好备选品。

常用单体汇总：

ethylene	英 /ˈeθɪliːn/ 美 /ˈeθɪliːn/	乙烯
propylene	英 /ˈprəʊpɪliːn/ 美 /ˈpropəlin/	丙烯
propene	英 /ˈprəʊpiːn/ 美 /ˈproˌpin/	
styrene	英 /ˈstaɪriːn/ 美 /ˈstaɪrin/	苯乙烯
vinylchloride	美 /ˈvaɪnl ˈklɔːraɪd/	氯乙烯
acrylonitrile	英 /ˌækrɪlə(ʊ)ˈnaɪtraɪl/ 美 /ˌækrəloʊˈnaɪtrɪl/	丙烯腈
acrylic acid	美 /əˈkrɪlɪk ˈæsɪd/	丙烯酸
vinylacetate	美 /ˈvaɪnl ˈæsɪteɪt/	乙酸乙烯酯
acrylamide	英 /əˈkrɪləmaɪd/ 美 /əˈkrɪləmaɪd/	丙烯酰胺
methyl acrylate		丙烯酸甲酯
methyl methacrylate		甲基丙烯酸甲酯

*提示：我们可以注意到很多单体都以"ene"为词尾。"ene"是烯烃的特征词尾。同理，"ane"为烷烃的词尾，"yl"为基团的词尾，"yne"为炔烃的词尾。例如 butene，pentene，hexene 的中文词意分别为丁烯、戊烯和己烯；butane，pentane，hexane 的中文词意分别为丁烷、戊烷和己烷。另外，与苯相关的词汇，大部分也以"ene"为词尾，如 benzene（苯）、toluene（甲苯）、styrene（苯乙烯）、pyrene（芘）、naphthalene（萘）等。

拓展阅读

Monomer

In chemistry, a monomer is a molecule that can react together with other monomer molecules to form a larger polymer chain or three dimensional network. Such a process is known as polymerization. Monomers can be classified into two classes according to different rules, such as natural vs synthetic monomers, cyclic vs linear monomers, polar vs nonpolar monomers, etc. The polymerization of one kind of monomer gives a homopolymer. When derived from two or more different monomers, the obtained polymer should be called copolymer. Usually, monomers are small molecules with low molecular weight, while sometimes also used to describe molecules with high molecular weight, e.g., the term of monomeric protein is one kind of protein that is able to combine to form multi-protein complexes. Oligomers are polymers consisting of a small number (typically under 100) of monomer subunits. Biopolymers are polymers consisting of natural monomers. Some famous polymeric products are coming from synthetic monomers with simple structures. For example, ethylene ($H_2C=CH_2$) is the monomer for polyethylene (PE). Vinyl chloride ($H_2C=CHCl$) is the monomer that leads to polyvinyl chloride (PVC). Styrene ($C_6H_5CH=CH_2$) is the monomer for polystyrene (PS). Synthetic rubbers are often based on butadiene.

Unit 6
Solvent

- 读音：英 /ˈsɒlvənt/ 美 /ˈsɑːlvənt/
- 释义：n. 溶剂
- 英英释义：A liquid that can dissolve other substances (《柯林斯英汉双解大词典》).

派生词及相关词汇：

solution	英 /səˈluːʃn/ 美 /səˈluːʃn/	n.	溶液
dissolve	英 /dɪˈzɒlv/ 美 /dɪˈzɑːlv/	v.	溶解
dissolution	英 /ˌdɪsəˈluːʃn/ 美 /ˌdɪsəˈluːʃn/	n.	溶解
emulsion	英 /ɪˈmʌlʃn/ 美 /ɪˈmʌlʃn/	n.	乳液
swelling	英 /ˈswelɪŋ/ 美 /ˈswelɪŋ/	n.	溶胀
swell	英 /swel/ 美 /swel/	v.	溶胀
medium	英 /ˈmiːdiəm/ 美 /ˈmiːdiəm/	n.	介质
suspension	英 /səˈspenʃn/ 美 /səˈspenʃn/	n.	悬浮液
nanosuspension	英 /næˈnəʊsəspenʃn/ 美 /næˈnəʊsəspenʃn/	n.	纳米悬浮液
interface	英 /ˈɪntəfeɪs/ 美 /ˈɪntərfeɪs/	n.	界面
insoluble	英 /ɪnˈsɒljəb(ə)l/ 美 /ɪnˈsɑːljəb(ə)l/	adj.	不溶解的
soluble	英 /ˈsɒljəb(ə)l/ 美 /ˈsɑːljəb(ə)l/	adj.	可溶的

固定搭配/常用短语：

organic solvent	有机溶剂
polar solvent	极性溶剂
developing solvent	展开剂

例句

- Herein, we propose a facile approach to prepare a novel W/O emulsion system, in which ethyl acetate was used as oil phase (Shi J, *et al.*, *Composites Science and Technology*, 2022, 230: 109760).
  在此，我们提出了一种简单的方法来制备新型 W/O 乳液体系，其中乙酸乙酯用作油相。

- The obtained polymer is soluble in most of organic solvents, such as methylene chloride, chloroform, tetrahydrofuran and dimethylsulfoxide (Ren X, *et al.*, *Macromolecular Chemistry*

and Physics, 2019, 9: 1900044).
所得聚合物可溶于大多数有机溶剂，如二氯甲烷、氯仿、四氢呋喃和二甲基亚砜。

- The prepared nanoparticle cannot be dissolved into the commonly-used solvents, but can be afford a stable nanosuspension in water phase (Zhang H, *et al.*, *Chemical Communications*, 2019, 55: 1040-1043).
制备的纳米颗粒不能溶解在常用溶剂中，但可以在水相中形成稳定的纳米悬浮液。

常用溶剂汇总：

pentane	英 /'penteɪn/ 美 /'penteɪn/	戊烷
petroleum ether		石油醚
hexane	英 /'hekseɪn/ 美 /'hek,seɪn/	己烷
cyclohexane	英 /ˌsaɪkləʊ'hekseɪn/ 美 /ˌsaɪkloʊ'hek,seɪn/	环己烷
cyclopentane	英 /ˌsaɪkləʊ'penteɪn/ 美 /ˌsaɪklə'pen,teɪn/	环戊烷
heptane	英 /'hepteɪn/ 美 /'hepteɪn/	庚烷
methylene chloride		二氯甲烷
chloroform	英 /'klɒrəfɔːm/ 美 /'klɔːrəfɔːrm/	氯仿
butyl chloride		丁基氯
carbon tetrachloride		四氯化碳
benzene	英 /'benziːn/ 美 /'benziːn/	苯
toluene	英 /'tɒljʊˌiːn/ 美 /'taljʊˌin/	甲苯
p-xylene		对二甲苯
chlorobenzene	英 /ˌklɔːrəʊ'benziːn/ 美 /ˌkloʊroʊ'ben,ziːn/	氯苯
o-dichlorobenzene		邻二氯苯
methanol	英 /'meθənɒl/ 美 /'meθənɔːl/	甲醇
ethanol	英 /'eθənɒl/ 美 /'eθənɔːl/	乙醇
propanol	英 /'prəʊpənɒl/ 美 /'prɑːpənɑːl/	丙醇
butanol	英 /'bjuːtəˌnɒl/ 美 /'bjutənol/	丁醇
ethyl ether		乙醚
tetrahydrofuran	英 /ˌtetrəˌhaɪdrə'fjʊəræn/ 美 /tetrəhaɪdrə'fjʊərən/	四氢呋喃
acetone	英 /'æsɪtəʊn/ 美 /'æsɪtoʊn/	丙酮
acetic acid		乙酸
ethyl acetate		乙酸乙酯
trifluoroacetic acid		三氟乙酸
pyridine	英 /'pɪrɪdiːn/ 美 /'pɪrɪˌdin/	吡啶
dimethyl formamide		二甲基甲酰胺
dimethyl sulfoxide		二甲基亚砜

Question (Please answer in English):
Which solvents are usually used as reaction solvents in polymerizations?

拓展阅读

Tu Youyou and Ethyl Ether

Tu Youyou is a Chinese pharmaceutical chemist and malariologist. She discovered artemisinin (青蒿素), which can be used to treat malaria (疟疾). It is a breakthrough in twentieth-century tropical medicine, saving millions of lives in South China, Southeast Asia, Africa, and South America. For her work, Tu Youyou received the 2015 Nobel Prize in Physiology or Medicine.

She experienced a lot of failures in her experiments in the early stage. After then, Tu Youyou realized that temperature was the key factor in extracting effective anti-malarial ingredients from herbs. She redesigned a new extraction strategy by using ethyl ether as the solvent because ethyl ether has a low boiling point which can lower the adverse effect of high temperature in extracting process. The container was filled with ethyl ether and the *Artemisia annua L.* (青蒿) was immersed in it to extract test samples. The results showed that after removing the acidic part of *Artemisia annua L.* ethyl ether extract, the remaining neutral part had the best efficacy. After hundreds of attempts, a preliminary breakthrough was made. Experiments confirmed that the inhibition rate of *Artemisia annua L.* ethyl ether neutral extract on rodent plasmodium (鼠疟原虫) reached 100 percent. Since ethyl ether is a type of organic solvent and viewed as a harmful chemical, Tu Youyou volunteered to be the first human test subject. "As head of this research group, I had the responsibility," she said. It was safe, so she conducted successful clinical trials with human patients.

Unit 7
Purification

- 读音：英 /ˌpjʊərɪfɪˈkeɪʃ(ə)n/ 美 /ˌpjʊrɪfɪˈkeɪʃ(ə)n/
- 释义：*n.* 纯化，提纯
- 英英释义：The act of removing useless substances from something.

同根词：

purify	英 /ˈpjʊərɪfaɪ/ 美 /ˈpjʊrɪfaɪ/	v.	纯化，提纯
purity	英 /ˈpjʊərəti/ 美 /ˈpjʊrəti/	n.	纯度
impurity	英 /ɪmˈpjʊərəti/ 美 /ɪmˈpjʊrəti/	n.	杂质

例 句

- In this study, the monomer was purified by using the column chromatography.
  本研究采用柱层析法对单体进行纯化。
- The purification procedure is very important in preparation of polymers with narrow range distribution.
  在窄分布聚合物的制备中，纯化过程是非常重要的。
- The purity of the drinking water is tested regularly.
  饮用水的纯度定期检测。

派生词及相关词汇：

conversion	英 /kənˈvɜːʃn/ 美 /kənˈvɜːrʒn/	n.	转化
yield	英 /jiːld/ 美 /jiːld/	n.	收率
distill	英 /dɪˈstɪl/ 美 /dɪˈstɪl/	v.	蒸馏
recrystallization	英 /riːˌkrɪstəlaɪˈzeɪʃən/ 美 /rekrɪstələˈzeɪʃn/	n.	重结晶
precipitate	英 /prɪˈsɪpɪteɪt/ 美 /prɪˈsɪpɪteɪt/	v.	沉淀
precipitation	英 /prɪˌsɪpɪˈteɪʃn/ 美 /prɪˌsɪpɪˈteɪʃn/	n.	沉淀
degas	英 /diːˈɡæs/ 美 /diˈɡæs/	v.	脱气
degassing	英 /diːˈɡæsɪŋ/ 美 /diˈɡæsɪŋ/	n.	脱气

（续表）

英文	音标	词性	中文
separate	英 /'seprət/　美 /'seprət/	v.	（使）分离
separation	英 /ˌsepə'reɪʃn/　美 /ˌsepə'reɪʃn/	n.	分离
removal	英 /rɪ'muːvl/　美 /rɪ'muːvl/	n.	移除
residue	英 /'rezɪdjuː/　美 /'rezɪduː/	n.	残渣
dialysis	英 /ˌdaɪ'æləsɪs/　美 /ˌdaɪ'æləsɪs/	n.	透析
elute	英 /ɪ'luːt/　美 /ɪ'ljut/	v.	洗提
chromatographic	英 /krəʊˌmætə'græfɪk/　美 /kroʊˌmætə'græfɪk/	adj.	色谱法的，色析法的
chromatographically	英 /krəʊmætə'græfɪkli/　美 /kroʊmætə'græfɪkli/	adv.	色析法地，色谱法地
centrifuge	英 /'sentrɪfjuːdʒ/　美 /'sentrɪfjuːdʒ/	n.	离心机
		v.	使……受离心作用
centrifugation	英 /sentrɪfjuː'geɪʃən/　美 /sentrɪfjuː'geɪʃn/	n.	离心分离
extract	英 /'ekstrækt/　美 /'ekstrækt/	v.	提取，萃取
extraction	英 /ɪk'strækʃn/　美 /ɪk'strækʃn/	n.	提取，萃取
percolation	英 /ˌpɜːkə'leɪʃn/　美 /ˌpɜːrkə'leɪʃn/	n.	浸透
filter	英 /'fɪltə(r)/　美 /'fɪltər/	v.	过滤
filtration	英 /fɪl'treɪʃn/　美 /fɪl'treɪʃn/	n.	过滤
selectivity	英 /səˌlek'tɪvəti/　美 /səˌlek'tɪvəti/	n.	选择性
exchange	英 /ɪks'tʃeɪndʒ/　美 /ɪks'tʃeɪndʒ/	n.	交换
		v.	交换
column	英 /'kɒləm/　美 /'kɑːləm/	n.	柱
layer	英 /'leɪə(r)/　美 /'leɪər/	n.	层
substance	英 /'sʌbstəns/　美 /'sʌbstəns/	n.	物质
analogue	英 /'ænəlɒg/　美 /'ænəlɔːg/	n.	类似物
derivative	英 /dɪ'rɪvətɪv/　美 /dɪ'rɪvətɪv/	n.	派生物

固定搭配/常用短语：

conversion rate	转化率
freeze drying	冷冻干燥
dialysis tube/bag	透析袋
chromatographic column	色谱柱
thin layer chromatography	薄层色谱法

Question (Please answer in English):
Which kind of hybrid solvents can be used in extraction procedure? And why?

拓展阅读

Chromatography and Its History

In chemical analysis, nowadays, chromatography is a commonly used laboratory technique for the separation of a mixture into its components. Chromatography was first devised by a Russian scientist Mikhail Tsvet in 1900. Initially, this technique was used to separate plant pigments. Since these components can be separated into bands of different colors (green, orange, and yellow, respectively), they directly entitled the name of the technique. New types of chromatography substantially developed in the 1940s and 1950s, which made the technique useful for many separation processes. For these works, Archer John Porter Martin and Richard Laurence Millington Synge won the 1952 Nobel Prize in Chemistry. They established the principles and basic techniques of partition chromatography, and their work encouraged the rapid development of several chromatographic methods: paper chromatography, gas chromatography (GC), and what would become known as high performance liquid chromatography (HPLC). Since then, the technology has advanced rapidly. Most important, one of the chromatography techniques, size exclusion chromatography has become a commonly used tool in polymer science.

Unit 8
Reaction

- 读音：英 /ri'ækʃn/ 美 /ri'ækʃn/
- 释义：*n.* 反应
- 英英释义：A chemical process in which two or more substances act mutually on each other and are changed into different substances, or one substance changes into two or more other substances (《新牛津英语词典》).

同根词：

react	英 /ri'ækt/ 美 /ri'ækt/	*v.*	反应
reactor	英 /ri'æktə(r)/ 美 /ri'æktər/	*n.*	反应器
reactant	英 /ri'æktənt/ 美 /ri'æktənt/	*n.*	反应物
reactive	英 /ri'æktɪv/ 美 /ri'æktɪv/	*adj.*	反应的
reactivity	英 /ˌri:æk'tɪvəti/ 美 /ˌri:æk'tɪvəti/	*n.*	反应活性

例 句

- Carboxyl groups can chemically react with hydroxyls to form ester bonds, even at low temperature.
  即使在低温下，羧基也能与羟基发生化学反应形成酯键。
- Intense heat was released in the reaction system.
  反应体系中释放出强热。
- Allyl-based monomers and acrylate-based monomers differ in chemical reactivity in radical polymerization.
  在自由基聚合中，烯丙基单体和丙烯酸酯基单体的化学反应活性不同。
- Conversion degree of reactant increased with the addition of Pd(0).
  反应物转化率随零价钯的加入而提高。

派生词及相关词汇：

flask	英 /flɑ:sk/ 美 /flæsk/	*n.*	烧瓶
catalyst	英 /'kætəlɪst/ 美 /'kætəlɪst/	*n.*	催化剂
catalyze	英 /'kætəlaɪz/ 美 /'kætəlaɪz/	*v.*	催化

Unit 8 Reaction

（续表）

reflux	英 /ˈriːflʌks/ 美 /ˈriːflʌks/	n.	回流
heat	英 /hiːt/ 美 /hiːt/	v.	加热
freeze	英 /friːz/ 美 /friːz/	v.	冷冻
stir	英 /stɜː(r)/ 美 /stɜːr/	v.	搅拌
shake	英 /ʃeɪk/ 美 /ʃeɪk/	v.	摇动
incubate	英 /ˈɪŋkjubeɪt/ 美 /ˈɪŋkjubeɪt/	v.	孵化，培育
residue	英 /ˈrezɪdjuː/ 美 /ˈrezɪduː/	n.	残渣
vacuum	英 /ˈvækjuːm/ 美 /ˈvækjuːm/	n.	真空

固定搭配/常用短语：

reactivity ratio	竞聚率
reaction time	反应时间
reaction temperature	反应温度
chemical reaction	化学反应
reaction mechanism	反应机理
reaction rate	反应速率
chain reaction	链式反应，连锁反应
reaction system	反应系统
reaction kinetics	反应动力学
side reaction	副反应
oxidation reaction	氧化反应
coupling reaction	偶联反应
tank reactor	反应釜
hydrothermal synthesis reactor	水热合成反应釜
intensely stir	剧烈搅拌
freeze drying	冷冻干燥
in vacuum	真空中

拓展阅读

Ziegler-Natta

A Ziegler-Natta catalyst, named after German Karl Ziegler and Italian Giulio Natta, is a catalyst used in the synthesis of polymers of 1-alkenes (alpha-olefins). They won the 1963 Nobel Prize in Chemistry. The polymerization using Ziegler-Natta catalyst to control the structure of the product can be named as Ziegler-Natta polymerization. Free radical vinyl polymerization can only give branched polyethylene, and propylene won't polymerize at all by free radical polymerization. Natta first used Ti-based catalysts polymerization to synthesize poly(propylene). He discovered that these polymers are crystalline materials and ascribed their crystallinity to a special feature of the polymer structure called stereoregularity (等规立构). In the 1970s, $MgCl_2$ was discovered to possess much higher activity of the Ti-based catalysts. These catalysts were so active that the removal of unwanted amorphous polymer and removal of residual titanium were no longer needed in the product. They enabled the commercialization of linear low density polyethylene resins and allowed the development of noncrystalline copolymers.

Unit 9
Characterization

- 读音：英 /ˌkærəktəraɪˈzeɪʃn/　美 /ˌkærəktərəˈzeɪʃn/
- 释义：*n.* 表征
- 英英释义：A description of the most typical or important peculiarities of something.

同根词：

character	英 /ˈkærəktə(r)/　美 /ˈkærəktər/	n.	特性
characterize	英 /ˈkærəktəraɪz/　美 /ˈkærəktəraɪz/	v.	表征，描述
characteristic	英 /ˌkærəktəˈrɪstɪk/　美 /ˌkerəktəˈrɪstɪk/	n. adj.	特征，特点 独特的，典型的
characterized	英 /ˈkærəktəˌraɪzd/　美 /ˈkærəktəraɪzd/	adj.	以……为特点的
characteristically	英 /ˌkærəktəˈrɪstɪkli/　美 /ˌkærəktəˈrɪstɪkli/	adv.	典型地

例 句

- We then developed a series of characterizations of the newly synthesized polymer-based nanocomposite.
  然后我们对新合成的聚合物基纳米复合材料进行了一系列的表征。
- The character of these boundary conditions is that of two coupled second-order ordinary differential equations.
  这些边界条件的性质是两个耦合的二阶常微分方程。
- The laboratory will conduct confirmatory tests and further characterize the virus.
  该实验室将进行确认检测，进一步确定病毒特征。
- On the other hand, fever is one of the characteristic features of the flu for all ages.
  另一方面，发烧是所有年龄段的人患流感的症状之一。
- Chemical laser is characteristically energized by an exoergic chemical reaction.
  化学激光器的特点是以释能化学反应提供能源。

固定搭配/常用短语：

general character	一般特征
main character	主要特征
characteristic equation	特性方程

（续表）

be characterized by	以……为特征
characterization factor	特性参数
material characterization	材料表征
morphological characterization	形貌表征

派生词及相关词汇：

obtain	英 /əb'teɪn/ 美 /əb'teɪn/	v.	获得，获取
perform	英 /pə'fɔːm/ 美 /pər'fɔːrm/	v.	执行
record	英 /'rekɔːd/ 美 /'rekərd/	v. n.	记录 记录
spectrum	英 /'spektrəm/ 美 /'spektrəm/	n.	光谱
spectra	英 /'spektrə/ 美 /'spektrə/	n.	光谱（spectrum 的复数形式）

例 句

- UV-visible absorption spectra were obtained on a Shimadzu UV-visible spectrometer model UV-2550.
  紫外-可见吸收光谱由 UV-2550 型岛津紫外-可见光谱仪获得。
- TGA was performed on PerkinElmer Pyris 6.
  热重分析在 PerkinElmer Pyris 6 上完成。
- Fluorescence spectra were recorded on a Shimadzu RF-5301PC.
  荧光光谱记录在岛津 RF-5301PC 上。

背景知识

characterization 是高分子领域英文文章中出现的高频词汇。很多文章均以 characterization 作为实验部分中的一个小标题。此外，characterization 更是高分子领域英文文章题目中使用的高频词，经常与 synthesis 一起使用（synthesis and characterization of sth），用于表示某种物质的合成和表征，例如：

- Synthesis and Characterization of Bio-based Alkyl Terminal Hyperbranched Polyglycerols: A Detailed Study of Their Plasticization Effect and Migration Resistance (Lee K W, et al., *Green Chemistry*, 2016, 18: 999-1009).
  生物基烷基末端超支化聚甘油的合成与表征:其塑化效应和迁移阻力的详细研究
- Synthesis and Characterization of 4-Vinylimidazolium/Styrene-Cografted Anion-Conducting Electrolyte Membranes (Hamada T, et al., *Macromolecular Chemistry and Physics*, 2021, 222: 2100028).
  4-乙烯基咪唑/苯乙烯共接枝阴离子导电电解质膜的合成与表征
- Synthesis and Characterization of Bisphthalonitrile-terminated Polyimide Precursors with Unique Advantages in Processing and Adhesive Properties (Liu C, et al., *Polymer*, 2021, 212: 123290).
  具有独特加工性能和粘接性能优势的端基双邻苯二甲腈聚酰亚胺前驱体的合成与表征

Unit 10
Processing

- 读音：英 /ˈprəʊsesɪŋ/　美 /ˈprɑːsesɪŋ/
- 释义：*n.* 加工处理，成型
- 英英释义：The series of actions that are taken to change raw materials during the production of goods (Cambridge Dictionary).

　　成型加工，是高分子由微观向宏观高分子材料迈进的重要环节，主要分为模压成型、挤出成型、压延成型和注射成型等类别。

◆ **compression molding** 模压成型

　　compression 英 /kəmˈpreʃn/　美 /kəmˈpreʃn/　*n.* 模压，压缩
　　英英释义：Compression molding is also called press molding, and compression molding is mainly used for molding thermosetting plastics such as phenolic resin, urea-formaldehyde resin, and unsaturated polyester resin.

例 句

　　The result shows that the biodegradable polymer materials can be used to produce compression molding products such as tableware for fast food and tray for packaging eggs.
　　结果表明，可降解高分子材料可用于生产快餐餐具和鸡蛋包装托盘等压制成型产品。

◆ **extrusion molding** 挤出成型

　　extrusion 英 /ɪkˈstruːʒn/　美 /ɪkˈstruːʒn/　*n.* 挤出
　　英英释义：Extrusion molding is a method of extruding a heated resin through a die using an extruder to extrude a desired shape of the article.

例 句

　　This paper briefly introduces the production technology and discusses the common problems of the extrusion molding of Nylon 66.
　　这篇文章主要介绍尼龙 66 的挤出成型工艺，并讨论了生产中易出现的问题。

◆ **calendering molding** 压延成型

　　calendering 英 /ˈkælɪndərɪŋ/　美 /kæˈlɪndərɪŋ/　*n.* 压延
　　英英释义：Calendering molding (Fig.1-3) is the process of processing (kneading, filtering, etc.)

the resin into a film or sheet by the gap between two or more opposite rolling rolls of a calender, and then peeling off from the calender roll.

例句

Calendering molding is a molding method mainly used for polyvinyl chloride resins, and can manufacture films, sheets, artificial leathers, floor tiles and the like.

压延成型是一种主要用于聚氯乙烯树脂的成型方法，可生产薄膜、片材、人造革、地砖等。

◆ **injection molding** 注射成型

injection 英 /ɪn'dʒekʃn/ 美 /ɪn'dʒekʃn/ n. 注射

英英释义：Injection molding is a method in which a thermoplastic melt is injected into a mold under high pressure using an injection molding machine (or an injection machine) to obtain a product by cooling and solidifying.

例句

Solid polyurethanes are similar to injection molding materials, and more suited to thinner wall sections.

固体聚氨酯类似注射成型材料，更适合生产片层材料。

Figure1-3　The products prepared via calendering molding

◆ **blow molding** 吹塑成型

blow 英 /bləʊ/ 美 /bloʊ/ v. 吹；n. 大风，强风

英英释义：Blow molding is a method in which a hot resin parison closed in a mold is inflated into a hollow article by the pressure of compressed air, and the blow molding includes two methods of blown film and blown hollow article.

例 句

The invention is an integrally blow-molded bag-in-container obtainable by blow-molding an injection molded multi-layer preform.

本发明是一种通过对注射成型的多层预制坯进行吹塑而获得的一体吹塑内装袋容器。

◆ **foaming molding** 发泡成型

foaming 英 /ˈfəʊmɪŋ/ 美 /ˈfoʊmɪŋ/ *adj.* 起泡的

例 句

The process of rotational foaming molding of liquid reactive resin was studied, with unsaturated polyester resin as the shell of unfoamed resin molded part, and polyurethane as foamed core layer.

以不饱和聚酯树脂作为非发泡树脂成型制品的外壳，用聚氨酯树脂发泡形成泡沫芯层，研究了液体反应树脂的旋转发泡成型的工艺过程。

派生词及相关词汇：

coating	英 /ˈkəʊtɪŋ/ 美 /ˈkoʊtɪŋ/	n.	涂层
shaping	英 /ˈʃeɪpɪŋ/ 美 /ˈʃeɪpɪŋ/	n.	成型
additive	英 /ˈædətɪv/ 美 /ˈædətɪv/	n.	添加剂
softener	英 /ˈsɒfnə(r)/ 美 /ˈsɔːfnər/	n.	软化剂
accelerator	英 /əkˈseləreɪtə(r)/ 美 /əkˈseləreɪtər/	n.	促进剂
mould	英 /məʊld/ 美 /moʊld/	n.	模具

Unit 11
Property

- 读音：英 /ˈprɒpəti/　美 /ˈprɑːpərti/
- 释义：*n.* 性能
- 英英释义：A quality in a substance or material, especially one that means that it can be used in a particular way (Cambridge Dictionary).

同（近）义词：

performance	英 /pəˈfɔːməns/　美 /pərˈfɔːrməns/	*n.*	性能，表现
peculiarity	英 /pɪˌkjuːliˈærəti/　美 /pɪˌkjuːliˈærəti/	*n.*	特点，特性
capability	英 /ˌkeɪpəˈbɪləti/　美 /ˌkeɪpəˈbɪləti/	*n.*	能力
ability	英 /əˈbɪləti/　美 /əˈbɪləti/	*n.*	能力
nature	英 /ˈneɪtʃə(r)/　美 /ˈneɪtʃər/	*n.*	本性

*提示: performance, peculiarity, capability, ability, nature 和 property 均可用于描述聚合物或者某种材料的性能、性质，但侧重点不同。Performance 一般指代的是材料在某方面的优异性能；peculiarity 则强调特有的性能，与众不同的、独特的性能，可用于表述优点，也可表述缺陷；capability 和 ability 意思相近，主要用于描述材料有能力达成某方面的作用，侧重于能力的描述；nature 则主要用于表述材料本身固有的性质，强调本性；property 在表述性能时具有普适性，各种性能描述一般均适用。

例 句

- Super alloy is a high-performance material which has extensively applied to aviation industries, and currently began to use in oil and chemical civil industry application.
  超合金是一种高性能的材料，已广泛应用于航空行业，目前正在石油、化工等民用工业中应用。

- The percolation behavior and positive vapor coefficient peculiarity were investigated, and the effect of different ways of thermal treatments on the PVC peculiarity was also explored (Liu Y, *et al.*, *Materials Research Innovations*, 2006, 10: 52-57).
  本文研究了聚氯乙烯的渗流行为和正蒸气系数特性，并探讨了不同热处理方式对聚氯乙烯特性的影响。

- The kind of polymers has shown the synthetic variety, the advanced capability and the wide applicability in contrast to the reported analogues (Tan J, *et al.*, *Chinese Chemical Letters*, 2016,

27: 1405-1411).
与已报道的类似物相比，这类聚合物表现出了合成的多样性、先进的性能和广泛的适用性。

- The synthesis and dynamic nature of macromolecular systems controlled by rotaxane macromolecular switches are introduced to discuss the significance of rotaxane linking of polymer chains and its topological switching (Takata T, *ACS Central Science*, 2020, 6: 129-143).
本文介绍了轮烷高分子开关控制的高分子体系的合成及其动力学性质，讨论了轮烷连接高分子链及其拓扑开关的意义。

- This paper scientifically discloses the importance of surface property of polymer particles on the rheological property of pc (Lu Z, *et al*., *Colloids and Surfaces A*: *Physicochemical and Engineering Aspects*, 2017, 520: 154-165).
本文科学地揭示了聚合物颗粒表面性质对聚碳酸酯流变性能的重要性。

固定搭配/常用短语：

mechanical property	力学性能
conformational property	构象性质
optical property	光学性质
electrical property	电性能
degradable property	可降解性能
thermal property	热性能（Fig.1-4）
glassy state	玻璃态
rubbery state	橡胶态
viscous flow state	黏流态
glass-transition temperature	玻璃化转变温度
melting temperature	熔融温度

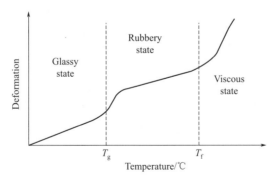

Figure1-4　Thermal property of polymer

背景知识

Property 是英文文章题目中使用的高频词，经常与 synthesis 或 characterization 一起使用（synthesis, characterization and property of sth），用于表示某种物质的合成、表征与性能研究，例如：

- Synthesis, Characterization, and Properties of Novel Phenylene-Silazane-Acetylene Polymers (Wang R, *et al.*, *Polymer*, 2010, 51: 5970-5976).
  新型亚苯-硅氮-乙炔聚合物的合成、表征及性能研究
- Conjugated Polymers Based on a S-and N-Containing Heteroarene: Synthesis, Characterization, and Semiconducting Properties (Chen Y, *et al.*, *Macromolecules*, 2011, 44: 5178-5185).
  基于含硫和含氮杂芳烃的共轭聚合物：合成、表征和半导体性能
- Nanoscopic Polymer Particles with a Well-defined Surface: Synthesis, Characterization, and Properties (Ballauff M, *Macromolecular Chemistry & Physics*, 2003, 204: 220-234).
  具有良好表面的纳米聚合物颗粒：合成、表征和性能

Unit 12
Degradation

- 读音：英 /ˌdegrəˈdeɪʃn/　美 /ˌdegrəˈdeɪʃn/
- 释义：*n.* 降解
- 英英释义：The process in which the quality of something is destroyed or spoiled (Cambridge Dictionary).

同根词：

degrade	英 /dɪˈgreɪd/　美 /dɪˈgreɪd/	*v.*	降解
degradable	英 /dɪˈgreɪdəbl/　美 /dɪˈgreɪdəbl/	*adj.*	可降解的
biodegradable	英 /ˌbaɪəʊdɪˈgreɪdəbl/　美 /ˌbaɪoʊdɪˈgreɪdəbl/	*adj.*	可生物降解的
degraded	英 /dɪˈgreɪdɪd/　美 /dɪˈgreɪdɪd/	*adj.*	已降解的

固定搭配/常用短语：

degradation mechanism	降解机制
thermal degradation	热降解
oxidative degradation	氧化降解
degradation product	降解产物
degradation rate	降解速率
degradable polymer	可降解的高分子
degradable plastic	可降解的塑料
biodegradable polymer	可生物降解的高分子

例 句

- The basic characteristics of degradable plastics and common-used plastics are analyzed and discussed.
  分析讨论了普通塑料和降解塑料的降解性能。
- The sun's UV rays can degrade resin on the outer walls.
  太阳的紫外线可以降解外墙的树脂。
- Base on the analysis of products, the degradation mechanism of PVC was deduced.
  在产物分析的基础上，推测了聚氯乙烯的降解机理。

- The results show that PS and PS/PP can be degraded into oily product in supercritical water.
  实验结果显示，超临界水可将聚苯乙烯及聚苯乙烯/聚丙烯混合塑料降解为液态油状物。

背景知识

在正常环境下，绝大多数聚合物会发生降解，降解过程一般较为缓慢。在热、超声、光照、水解作用等外界刺激下，降解速率有可能显著提高。此外，酸、碱等也可能加快降解速率。通过上述作用，聚合物被降解为低分子量产物、寡聚体、三聚体、二聚体等。

派生词及相关词汇：

ultraviolet	英 /ˌʌltrə'vaɪələt/ 美 /ˌʌltrə'vaɪələt/	n. adj.	紫外线 紫外的
ultrasonic	英 /ˌʌltrə'sɒnɪk/ 美 /ˌʌltrə'sɑːnɪk/	n. adj.	超声波 超声的
temperature	英 /'temprətʃə(r)/ 美 /'temprətʃər/	n.	温度
centigrade	英 /'sentɪɡreɪd/ 美 /'sentɪɡreɪd/	n. adj.	摄氏度（℃） 摄氏度的
hydrolysis	英 /haɪ'drɒlɪsɪs/ 美 /haɪ'drɑːlɪsɪs/	n.	水解作用
stimulate	英 /'stɪmjuleɪt/ 美 /'stɪmjuleɪt/	v.	刺激
stimulation	英 /ˌstɪmju'leɪʃn/ 美 /ˌstɪmju'leɪʃn/	n.	刺激
acid	英 /'æsɪd/ 美 /'æsɪd/	n.	酸
acidic	英 /ə'sɪdɪk/ 美 /ə'sɪdɪk/	adj.	酸性的
alkali	英 /'ælkəlaɪ/ 美 /'ælkəlaɪ/	n.	碱
alkaline	英 /'ælkəlaɪn/ 美 /'ælkəlaɪn/	adj.	碱性的
oligomer	英 /ɒ'lɪɡəmə(r)/ 美 /ə'lɪɡəmər/	n.	寡聚体，低聚物
trimer	英 /'traɪmə/ 美 /'traɪmər/	n.	三聚体
dimer	英 /'daɪmə/ 美 /'daɪmər/	n.	二聚体
fraction	英 /'frækʃn/ 美 /'frækʃn/	n.	部分

常见的可降解聚合物：

poly(3-hydroxybutyrate)	聚-3-羟基丁酸（酯）	PHB
polyhydroxyvalerate	聚羟基戊酸酯	PHV
polylactic acid	聚乳酸	PLA
polycaprolactone	聚己内酯	PCL
poly(butylene succinate)	聚丁二酸丁二醇酯	PBS
polyglycolic acid	聚乙醇酸	PGA
poly(lactic co glycolic acid)	聚乳酸羟基乙酸共聚物	PLGA

拓展阅读

Polylactic Acid: A Typical Degradable Polymer

Polylactic acid, also known as poly(lactic acid) or polylactide (PLA), can be obtained by condensation of lactic acid or ring-opening polymerization of lactide. PLA can range from amorphous glassy polymer to semi-crystalline and highly crystalline polymer with a glass transition temperature (T_g) of 60-65 ℃ and a melting temperature (T_m) of 130-180 ℃. PLA shows good solubility in water and commonly used organic solvents. Ethyl acetate is widely used due to its ease of access and low risk. It has been demonstrated that PLA objects can be fabricated by injection moulding, extrusion, solvent welding, machining and 3D printing. PLA can degrade into innocuous lactic acid under physiological conditions, making it suitable for use as degradable surgical implants. It breaks down inside the body from 6 months to 2 years, depending on the type used. Such a gradual degradation ability is desirable for a support structure. In 2021, PLA had the highest consumption volume of any bioplastic of the world.

Unit 13
Modulus

- 读音：英 /'mɒdjʊləs/　美 /'mɑdʒələs/
- 释义：*n.* 模量
- 英英释义：(1) A coefficient that expresses how much of a specified property is possessed by a specified substance（WordNet）；(2) A constant or coefficient that expresses usually numerically the degree to which a body or substance possesses a particular property (Merriam-Webster).

同根词：

| moduli | 英 /'mɒdjʊˌlaɪ/　美 /'mɑdʒəˌlaɪ/ | n. | 模（modulus 的复数形式） |

同（近）义词

| parameter | 英 /pə'ræmɪtə(r)/　美 /pə'ræmɪtər/ | n. | 参数 |
| coefficient | 英 /ˌkəʊɪ'fɪʃnt/　美 /ˌkoʊɪ'fɪʃnt/ | n. | 系数 |

固定搭配/常用短语：

Young modulus	杨氏模量
tensile modulus	拉伸模量
bulk modulus	体积模量
shear modulus	剪切模量
elastic modulus	弹性模量
flexural modulus	弯曲模量
storage modulus	储能模量
loss modulus	损耗模量
Poisson's ratio	泊松比

例 句

- However, it still remains a puzzling problem to apply these methods for directly measuring the elastic modulus of the polymer films supported by stiff substrates, mainly due to the convolution between the polymer and substrate moduli.
  然而，应用这些方法直接测量由刚性衬底支撑的聚合物薄膜的弹性模量仍然是一个棘手的问题，这主要是由于聚合物与衬底模量之间的卷积。

Unit 13 Modulus

- Storage modulus, loss modulus and damping behavior obtained from the DMA measurement was used to investigate the manner of nanotubes into and onto the three-layer polymer structure.
利用动态机械分析仪测试获得的存储模量、损耗模量和阻尼行为,研究了纳米管在三层聚合物结构中的进出方式。
- The addition of nanoparticles into a polymer matrix gave rise to approximately 50% decrease in the Young modulus, 70% decrease in strain at fracture, and 50% decrease in the storage modulus.
在聚合物基体中加入纳米颗粒后,杨氏模量降低了约 50%,断裂时应变降低了 70%,存储模量降低了 50%。

派生词及相关词汇:

strain	英 /streɪn/ 美 /streɪn/	n.	张力,拉力
compression	英 /kəm'preʃn/ 美 /kəm'preʃn/	n.	压缩
extension	英 /ɪk'stenʃn/ 美 /ɪk'stenʃn/	n.	拉伸
deformation	英 /ˌdiːfɔː'meɪʃn/ 美 /ˌdiːfɔːr'meɪʃn/	n.	形变
resistance	英 /rɪ'zɪstəns/ 美 /rɪ'zɪstəns/	n.	对抗
rigidity	英 /rɪ'dʒɪdəti/ 美 /rɪ'dʒɪdəti/	n.	硬度
stress	英 /stres/ 美 /stres/	n. v.	压力 使……受压
flexible	英 /'fleksəb(ə)l/ 美 /'fleksəb(ə)l/	adj.	柔韧的,灵活的
stiff	英 /stɪf/ 美 /stɪf/	adj.	僵硬的
stiffness	英 /'stɪfnəs/ 美 /'stɪfnəs/	n.	硬度
elastic	英 /ɪ'læstɪk/ 美 /ɪ'læstɪk/	adj.	弹性的
uniaxial	英 /juːnɪ'æksɪəl/ 美 /jʊnɪ'æksɪəl/	adj.	单轴的
perpendicular	英 /ˌpɜːpən'dɪkjələ(r)/ 美 /ˌpɜːrpən'dɪkjələr/	n. adj.	垂直 垂直的
orientation	英 /ˌɔːriən'teɪʃ(ə)n/ 美 /ˌɔːriən'teɪʃ(ə)n/	n.	方向
orientate	英 /'ɔːriənteɪt/ 美 /'ɔːriənteɪt/	v.	取向
orientated	英 /'ɔːriənteɪtɪd/ 美 /'ɔːriənteɪtɪd/	adj.	面向……的
transverse	英 /'trænzvɜːs/ 美 /'trænzvɜːrs/	adj.	横向的
longitudinal	英 /ˌlɒŋgɪ'tjuːdən(ə)l/ 美 /ˌlɑːndʒə'tuːdən(ə)l/	adj.	纵向的

拓展阅读

What Is Young Modulus?

Young modulus (*E* or *Y*) is a commonly-used parameter, which can be used to evaluate a solid's stiffness or resistance to elastic deformation under load. It is calculated based on the relationship between stress (force per unit area) and strain (proportional deformation). The basic principle is that a material undergoes elastic deformation when it is stretched or compressed, returning to its original shape when the load is removed. Generally, flexible material exhibits more deformation than that of stiff material. A low Young modulus value means the sample should be elastic, while a high Young modulus value means the solid is inelastic or stiff. The Young modulus can be calculated by the following equation.

$$E = \sigma / \varepsilon = (F/A) / (\Delta L/L_0) = FL_0 / A\Delta L$$

where E is Young modulus, usually expressed in pascal (Pa); σ and ε represent the uniaxial stress and strain, respectively; F is the force of compression or extension; A is the cross-sectional surface area or the cross-section perpendicular to the applied force; ΔL and L_0 represent the change in length (negative under compression; positive when stretched) and the original length, respectively.

Though the unit for Young modulus is Pa, values are most often expressed in terms of megapascal (MPa), gigapascals (GPa), newtons per square millimeter (N/mm^2), or kilo-newtons per square millimeter (kN/mm^2). The usual English unit is pounds per square inch (psi) or mega psi (Mpsi).

Young modulus describes tensile elasticity along a line when opposing forces are applied. It is the ratio of tensile stress to tensile strain. Besides Young modulus, there are still some other moduli can be used to measure elasticity.

The bulk modulus (K) is like Young modulus, except in three dimensions. It is a measure of volumetric elasticity, calculated as volumetric stress divided by volumetric strain.

The shear or modulus of rigidity (G) describes shear when an object is acted upon by opposing forces. It is calculated as shear stress over shear strain.

The axial modulus, P-wave modulus, and Lamé's first parameter are other moduli of elasticity. Poisson's ratio may be used to compare the transverse contraction strain to the longitudinal extension strain. Together with Hooke's law, these values describe the elastic properties of a solid material, sometimes are also applicable to semi solid materials.

Unit 14
Morphology

- 读音：英 /mɔː'fɒlədʒi/　美 /mɔːr'fɑːlədʒi/
- 释义：*n.* 形貌
- 词源：morpho-形状，形态，-logy 学说。
- 英英释义：The study of the structure or form of substances.

同根词：

morphological	英 /ˌmɔːfə'lɒdʒɪkl/　美 /ˌmɔːrfə'lɑːdʒɪkl/	*adj.*	形貌的
morphologically	英 /ˌmɔːfə'lɒdʒɪkli/　美 /mɔrfə'lɒdʒɪklɪ/	*adv.*	形貌地
micromorphology	英 /ˌmaɪkrəʊmɔː'fɒlədʒi/ 美 /ˌmaikrəumɔː'fɔlədʒi/	*n.*	微观形貌
micromorphological	英 /'maɪkrəʊˌmɔːfə'lɒdʒɪkəl/	*adj.*	微观形貌的
micromorphologically	英 /'maɪkrəʊˌmɔːfə'lɒdʒɪkəlɪ/	*adv.*	微观形貌地

例 句

- The transition of crystalline morphology is revealed in poly(*E*-caprolactone) (PCL) thin films as the polymer film thickness changes from hundreds of nanometers to several nanometers.
随着聚合物膜厚度从几百纳米到几纳米的变化，聚己内酯(PCL)薄膜的晶体形貌发生了转变。

- The interrelations between the morphological structure of inter-polymer complexes of sodium lignosulfonate (alginate) with chitosan and the structure of the initial biopolyelectrolytes was established (Brovko O S, *et al.*, *Macromolecular Research*, 2015, 23: 1059-1067).
建立了木质素磺酸钠(海藻酸钠)与壳聚糖聚合物间复合物的形貌结构与初始生物电解质结构之间的相互关系。

- Its micromorphology directly affects its water absorption mechanism and absorption characteristics (Li Y, *et al.*, *Journal of Applied Polymer Science*, 2009, 113: 3510-3519).
其微观形貌直接影响其吸水机理和吸水特性。

固定搭配/常用短语：

morphological character	形貌特征
surface morphology	表面形貌

（续表）

fracture surface morphology	断口形貌
crystal morphology	晶体形貌
two-dimension (2D)	二维
three-dimension (3D)	三维
hydrodynamic radius	流体动力学半径
mean square radius of gyration	均方回转半径
transmission electron microscopy	透射电子显微镜
scanning electron microscopy	扫描电子显微镜
atomic force microscope	原子力显微镜

派生词及相关词汇：

pellet	英 /ˈpelɪt/　美 /ˈpelɪt/	n.	小球
fiber	英 /ˈfaɪbə(r)/　美 /ˈfaɪbər/	n.	纤维
rod	英 /rɒd/　美 /rɑːd/	n.	棒
pore	英 /pɔː(r)/　美 /pɔːr/	n.	孔
microporous	英 /ˈmaɪkrəʊˈpɔːrəs/　美 /ˈmaɪkroʊˈpɔːrəs/	adj.	多微孔的
microporosity	英 /ˌmaɪkrəʊpɔːˈrɒsɪti/　美 /ˌmaɪkroʊpɔːˈrɑːsəti/	n.	孔隙率
resolution	英 /ˌrezəˈluːʃ(ə)n/　美 /ˌrezəˈluːʃ(ə)n/	n.	分辨率
dimension	英 /daɪˈmenʃn/　美 /daɪˈmenʃn/	n.	维度
particle	英 /ˈpɑːtɪkl/　美 /ˈpɑːrtɪkl/	n.	颗粒
chain	英 /tʃeɪn/　美 /tʃeɪn/	n.	（分子）链
fracture	英 /ˈfræktʃə(r)/　美 /ˈfræktʃər/	n.	断面，截面
radius	英 /ˈreɪdiəs/　美 /ˈreɪdiəs/	n.	半径
sphere	英 /sfɪə(r)/　美 /sfɪr/	n.	球
rough	英 /rʌf/　美 /rʌf/	adj.	粗糙的
smooth	英 /smuːð/　美 /smuːð/	adj.	平滑的
uniform	英 /ˈjuːnɪfɔːm/　美 /ˈjuːnɪfɔːrm/	adj.	均一的
matrix	英 /ˈmeɪtrɪks/　美 /ˈmeɪtrɪks/	n.	基质
matrices	英 /ˈmeɪtrɪsiːz/　美 /ˈmeɪtrɪsiːz/	n.	基质（matrix的复数形式）

拓展阅读

SEM, TEM and AFM

A scanning electron microscope (SEM) is a type of electron microscope that produces images of a sample by scanning the surface with a high energy beam of electrons in a raster scan pattern. The electron beam is scanned in a raster scan pattern, and the position of the beam is combined with the intensity of the detected signal to produce an image. The electrons interact with atoms in the sample, producing various signals that contain information about the surface morphology and elemental composition of the sample.

Transmission electron microscopy (TEM) is also electron microscopy technique with the same principles as SEM. In TEM technique, a beam of electrons is transmitted through a specimen to form an image. The specimen is most often an ultrathin section less than 100nm thick or a suspension on a grid. An image is formed from the interaction of the electrons with the sample as the beam is transmitted through the specimen. The image is then magnified and focused onto an imaging device, such as a fluorescent screen, a layer of photographic film, or a sensor such as a scintillator attached to a charge-coupled device. TEM technique is useful for microstructural analysis in biopolymer membranes and the substance with hollows or cavities, which are not visible by the SEM technique.

Atomic force microscopy (AFM) is a type of scanning probe microscopy (SPM). AFM can be used for force measurement, topographic imaging and manipulation. In force measurement, AFM can be used to measure the forces between the probe and the sample as a function of their mutual separation. For imaging, the reaction of the probe to the force that the sample imposes on it can be used to form an image of the three dimensional shape of a sample surface at a high resolution. In manipulation, the force between probe tip and sample can also be used to change the properties of the sample in a controlled way.

Unit 15
Element

- 读音：英 /ˈelɪmənt/ 美 /ˈelɪmənt/
- 释义：*n.* 元素
- 英英释义：Any of the more than 100 known substances (of which 92 occur naturally) that cannot be separated into simpler substances and that singly or in combination constitute all matter（Cambridge Dictionary）.

固定搭配/常用短语：

chemical element	化学元素
trace element	微量元素
metallic element	金属元素
rare earth element	稀土元素
periodic table of elements	元素周期表

高分子化学中常用的元素：

hydrogen	英 /ˈhaɪdrədʒən/ 美 /ˈhaɪdrədʒən/	*n.*	氢
lithium	英 /ˈlɪθiəm/ 美 /ˈlɪθiəm/	*n.*	锂
boron	英 /ˈbɔːrɒn/ 美 /ˈbɔːrɑːn/	*n.*	硼
carbon	英 /ˈkɑːbən/ 美 /ˈkɑːrbən/	*n.*	碳
nitrogen	英 /ˈnaɪtrədʒən/ 美 /ˈnaɪtrədʒən/	*n.*	氮
oxygen	英 /ˈɒksɪdʒən/ 美 /ˈɑːksɪdʒən/	*n.*	氧
fluorine	英 /ˈflɔːriːn/ 美 /ˈflɔːriːn/	*n.*	氟
sodium	英 /ˈsəʊdiəm/ 美 /ˈsoʊdiəm/	*n.*	钠
aluminum	英 /əˈluːmɪnəm/ 美 /əˈluːmɪnəm/	*n.*	铝
silicon	英 /ˈsɪlɪkən/ 美 /ˈsɪlɪkən/	*n.*	硅
phosphorus	英 /ˈfɒsfərəs/ 美 /ˈfɑːsfərəs/	*n.*	磷
sulfur	英 /ˈsʌlfə(r)/ 美 /ˈsʌlfər/	*n.*	硫
chlorine	英 /ˈklɔːriːn/ 美 /ˈklɔːriːn/	*n.*	氯
potassium	英 /pəˈtæsiəm/ 美 /pəˈtæsiəm/	*n.*	钾
calcium	英 /ˈkælsiəm/ 美 /ˈkælsiəm/	*n.*	钙

			（续表）
iron	英 /ˈaɪən/ 美 /ˈaɪərn/	*n.*	铁
manganese	英 /ˈmæŋɡəniːz/ 美 /ˈmæŋɡəniːz/	*n.*	锰

例 句

- The weight-average molecular weight and polydispersity index of a polystyrene sample initiated by *n*-butyl lithium was determined by high performance size exclusion chromatography on refractive index detector.
  采用配置示差检测器的高效尺寸排阻色谱测定了正丁基锂引发聚苯乙烯样品的重均分子量和多分散性指数。
- A molecule of water consists of two atoms of hydrogen and one atom of oxygen.
  水分子由两个氢原子和一个氧原子构成。
- The composite was prepared by mixing the polymer with commercially available carbon nanotubes in a solvent.
  该复合材料是将聚合物与市售碳纳米管混合在溶剂中制备的。
- The treatment of fiber under anhydrous conditions with potassium *tert*-butoxide solution is studied.
  研究了叔丁醇钾溶液在无水条件下对纤维的处理。
- Boron trifluoride is a commonly-used cationic initiator in ionic polymerization.
  三氟化硼是离子聚合中常用的阳离子引发剂。

拓展阅读

Periodic Table of Elements and Zhu Yuanzhang

In 1869, Xu Shou (徐寿, the pioneer of modern Chinese chemistry) learned of Mendeleev's periodic table and admired it so much that he hoped to introduce it to China. However, he encountered difficulties in the process of translation. Most of these more than 100 elements have no corresponding words in Chinese character. How to properly translate them into Chinese bothered him for a long time, until he came across the family tree of the Ming Dynasty. The founder of Ming Dynasty, Zhu Yuanzhang(朱元璋), specifically formulated the naming rules for later generations. The name should meet the principle of the "five elements"（五行）: gold (金), wood (木), water (水), fire (火), and earth (土). In the late Ming Dynasty, there were no words available to name the newly-born babies. So, they had to make new words. In this process, a large number of characters with "five elements" are produced. When Xu Shou saw these characters in the names in genealogy, he was very excited and read it carefully. He selected a large number of characters from it and was inspired to transform a number of characters. For examples, the words alkane (烷), alkene (烯), alkynes (炔) and hydrocarbon (烃) may come from the names Zhu Qinwan (朱勤烷), Zhu Xiyue (朱烯悦), Zhu Yongque (朱颙炔) and Zhu Qiongting (朱琼烃). It is so wonderful that the existence of the connection between Zhu Yuanzhang and the periodic table of the elements found by Mendeleev.

Unit 16
Function

- 读音：英 /ˈfʌŋkʃn/ 美 /ˈfʌŋkʃn/
- 释义：*n.* 功能，函数
- 英英释义：Something or someone is the useful thing that they do or are able to do; a relation or expression involving one or more variables（《新牛津英语词典》）.

同根词：

functional	英 /ˈfʌŋkʃənl/ 美 /ˈfʌŋkʃənl/	*adj.*	功能的
functionally	英 /ˈfʌŋkʃənəli/ 美 /ˈfʌŋkʃənəli/	*adv.*	功能地
functionality	英 /ˌfʌŋkʃəˈnæləti/ 美 /ˌfʌŋkʃəˈnæləti/	*n.*	功能
difunctional	英 /daɪˈfʌŋkʃənəl/ 美 /daɪˈfʌŋkʃənəl/	*adj.*	双官能团的
monofunctional	英 /mɒnəʊˈfʌŋkʃənəl/ 美 /ˌmɑnəʊˈfʌŋkʃənəl/	*adj.*	单官能团的
multifunctional	英 /ˌmʌltiˈfʌŋkʃənl/ 美 /ˌmʌltiˈfʌŋkʃənl/	*adj.*	多官能团的
polyfunctional	英 /ˌpɒlɪˈfʌŋkʃənəl/ 美 /ˌpɑːlɪˈfʌŋkʃənəl/	*adj.*	多官能团的

固定搭配/常用短语：

| density functional theory (DFT) | 密度泛函理论 |
| functional group | 官能团 |

例 句

- **Density functional theory** is computational technique used to predict the properties of molecules and bulk materials. It is a method for investigating the electronic structure of many-body systems and is based on a determination of a given system's electron density rather than its wavefunction (Nature portfolio, Springer).
  密度泛函理论是一种用于预测分子和块体材料性质的计算技术。它是研究多体系统的电子结构的一种方法，其基于确定给定系统的电子密度而不是其波函数。

- **Functional group**, any of numerous combinations of atoms that form parts of chemical molecules, that undergo characteristic reactions themselves, and that in many cases influence the reactivity of the remainder of each molecule.
  官能团是构成化学分子组成部分的众多原子组合中的任何一种，它们本身进行特征反应，并在许多情况下影响每个分子其余部分的反应性。

Unit 16 Function

常用的官能团：

hydroxyl	英 /haɪˈdrɒksaɪl/ 美 /haɪˈdrɑːksɪl/	n.	羟基
hydroxy	英 /haɪˈdrɒksi/ 美 /haɪˈdrɑːksi/	adj.	羟基的
carboxyl	英 /kɑːˈbɒksɪl/ 美 /kɑrˈbɑksɪl/	n.	羧基
methyl	英 /ˈmeθɪl/ 美 /ˈmeθəl/	n.	甲基
methylene	英 /ˈmeθɪˌliːn/ 美 /ˈmeθɪliːn/	n.	亚甲基
ethyl	英 /ˈeθɪl/ 美 /ˈeθɪl/	n.	乙基
propyl	英 /ˈprəʊpɪl/ 美 /ˈproʊpəl/	n.	丙基
butyl	英 /ˈbjuːtɪl/ 美 /ˈbjuːtəl/	n.	丁基
carbonyl	英 /ˈkɑːbəˌnaɪl/ 美 /ˈkɑrbənɪl/	n.	羰基
isocyanate	英 /ˌaɪsəʊˈsaɪəneɪt/ 美 /ˌaɪsəˈsaɪəˌneɪt/	n.	异氰酸酯
formyl	英 /ˈfɔːmaɪl/ 美 /ˈfɔrˌmɪl/	n.	甲酰基
amino	英 /əˈmiːnəʊ/ 美 /əˌmiːnoʊ/	n.	氨基
vinyl	英 /ˈvaɪnl/ 美 /ˈvaɪnl/	n.	乙烯基
allyl	英 /ˈælaɪl/ 美 /ˈælɪl/	n.	烯丙基
acrylate	英 /ˈækrɪˌleɪt/ 美 /ˈækrəˌleɪt/	n.	丙烯酸酯
phenyl	英 /ˈfiːnaɪl/ 美 /ˈfenəl/	n.	苯基

拓展阅读

Functional Groups

In organic chemistry, a functional group is a substituent or moiety in a molecule that contribute to the molecule's specific reactivity patterns and characteristic chemical reactions. This enables the design of chemical synthesis and precise prediction of physicochemical property of chemical compounds. Therefore, functional groups are a significant organizing feature of synthetic chemistry. For repeating units of polymers, functional groups attach to their nonpolar core of carbon atoms and thus add chemical character to carbon chains. Functional groups can also be charged, e.g. carboxylate salts (—COO—), which turns the molecule into a polyatomic ion or a complex ion.

Carbon-to-carbon and carbon-to-hydrogen bonds are extremely strong and the charge of the electrons in these covalent bonds is spread more or less evenly over the bonded atoms. So, hydrocarbons that contain only single bonds of these two types are not very reactive. The reactivity of a molecule can increase if it contains one or more weak bonds. For example, a vinyl group is a functional group with the formula —CH=CH_2, which contains the ethylene molecule (H_2C=CH_2) with one fewer hydrogen atom. Vinyl groups are able to polymerize with the aid of a radical initiator, forming chain polymers. An industrially important example is styrene, precursor to polystyrene (PS).

Unit 17
Instrument

- 读音：英 /ˈɪnstrəmənt/　美 /ˈɪnstrəmənt/
- 释义：*n.* 仪器，设备
- 英英释义：A tool or device used for a particular task, especially for delicate or scientific work.

同（近）义词：

facility	英 /fəˈsɪləti/　美 /fəˈsɪləti/	*n.*	设施，设备
equipment	英 /ɪˈkwɪpmənt/　美 /ɪˈkwɪpmənt/	*n.*	装备，设备
vehicle	英 /ˈviːəkl/　美 /ˈviːəkl/	*n.*	工具
implement	英 /ˈɪmplɪmənt/　美 /ˈɪmplɪmənt/	*n.*	工具，器具

派生词及相关词汇：

record	英 /ˈrekɔːd/　美 /ˈrekərd/	*n.*	记录，记载
		v.	记录，记载
perform	英 /pəˈfɔːm/　美 /pərˈfɔːrm/	*v.*	完成，表现
conduct	英 /kənˈdʌkt/　美 /kənˈdʌkt/	*v.*	实施，进行
sample	英 /ˈsɑːmpl/　美 /ˈsæmpl/	*n.*	样品
		v.	取样
obtain	英 /əbˈteɪn/　美 /əbˈteɪn/	*v.*	获得
reference	英 /ˈrefrəns/　美 /ˈrefrəns/	*n.*	参比
parameter	英 /pəˈræmɪtə(r)/　美 /pəˈræmɪtər/	*n.*	参数
data	英 /ˈdeɪtə/　美 /ˈdeɪtə/	*n.*	数据
laser	英 /ˈleɪzə(r)/　美 /ˈleɪzər/	*n.*	激光
ultraviolet	英 /ˌʌltrəˈvaɪələt/　美 /ˌʌltrəˈvaɪələt/	*n.*	紫外线
		adj.	紫外的
infrared	英 /ˌɪnfrəˈred/　美 /ˌɪnfrəˈred/	*n.*	红外线
		adj.	红外的
fluorescence	英 /fləˈresns/　美 /fləˈresns/	*n.*	荧光
beam	英 /biːm/　美 /biːm/	*n.*	束

Unit 17　Instrument

高分子化学中常用的仪器：

名称	释义	缩写	测试参数
Gel permeation chromatography	凝胶渗透色谱	GPC	Molecular weight
Nuclear magnetic resonance	核磁共振	NMR	Structural information
Thermal gravimetric analyzer	热重分析仪	TGA	Thermal properties
Fourier transform infrared spectrometer	傅里叶变换红外光谱仪	FTIR	Functional group
X-ray diffraction	X射线衍射	XRD	Crystals
X-ray photoelectron spectroscopy	X射线光电子能谱[学]	XPS	Elements and valence state
Differential scanning calorimetry	差示扫描量热法	DSC	Thermal properties, glass-transition
Dynamic mechanical analysis	动力学分析	DMA	Mechanical properties
Dynamic light scattering	动态光散射	DLS	Size
Ultraviolet and visible spectrophotometer	紫外可见分光光度计	UV/Vis	Optical property
Fluorescence spectrophotometer	荧光分光光度计	—	Optical property

例句

- TGA was performed on Perkin-Elmer Pyris 6 under a nitrogen flow. Accurately weighted amounts of samples were heated at a scanning rate of 10℃/min from 40 to 800 ℃ (Zhang H, *et al.*, *ACS Applied Materials & Interfaces*, 2015, 7: 23805-23811).
  热重分析在珀金埃尔默公司 Pyris 6 型仪器上进行，并在氮气保护下开展测试。精确称量的样品在 10℃/min 的扫描速率下由 40℃升至 800℃。

- FTIR spectra were recorded in the region of 400-4000cm^{-1} for each sample on a Varian-640 spectrophotometer. Samples were previously grounded and mixed thoroughly with KBr (Zhang H, *et al.*, *ACS Applied Materials & Interfaces*, 2015, 7: 23805-23811).
  在 Varian-640 分光光度计上记录每个样品在 400～4000cm^{-1} 范围内的红外光谱图。样品预先研磨并与溴化钾充分混合。

- The molecular weight, polydispersity and intrinsic viscosity were determined by Viscotek 270-doul detector-size exclusion chromatography which equipped with differential refractive index, viscometer, a Waters gel permeation chromatographic column and two-angle light scattering triplet detectors (Liu F, *et al.*, *Polymer Chemistry*, 2018, 9: 5024-5031).
  分子量、多分散性和特性黏度由 Viscotek 270 双检测器-尺寸排除色谱仪测定，该色谱配有差示折射率、黏度计、Waters 凝胶渗透色谱柱和两角光散射三重检测器。

- The dynamic light scattering measurement was performed by a commercialized spectrometer from Brookhaven Instrument Corporation. A He-Ne laser operating at 633nm was used as the light source (Liu F, *et al.*, *Polymer Chemistry*, 2018, 9: 5024-5031).
  动态光散射测量使用 Brookhaven 仪器公司的商品化光谱仪。采用氦-氖激光器 633nm 的光线作为光源。

Unit 18
Natural

- 读音：英 /'nætʃ(ə)rəl/　美 /'nætʃ(ə)rəl/
- 释义：*adj.* 天然的
- 英英释义：Things exist or occur in nature and are not made or caused by people.

固定搭配/常用短语：

natural polymer	天然高分子
non-natural polymer	非天然高分子
natural polymer foam	天然高分子泡沫材料
natural polymer fibres	天然聚合纤维
cationic natural polymer	阳离子天然大分子
natural polymer derivative	天然高分子衍生物
modified natural polymer	改性天然高分子
carbohydrate polymer	碳水化合物聚合物

例 句

- To compare with synthetic polymer, natural polymers usually show good biocompatibility and low cytotoxicity.
  与合成高分子相比，天然高分子通常具有较好的生物相容性和低毒性。
- The study and application status using natural polymer fibres was introduced in this paper and a brief evaluation was given.
  本文介绍了近年来天然聚合纤维的研究和应用情况，并作了简要的评述。

常用的天然聚合物：

cellulose	英 /'seljuləʊs/　美 /'seljuloʊs/	*n.*	纤维素
starch	英 /stɑːtʃ/　美 /stɑːrtʃ/	*n.*	淀粉
amylose	英 /'æmɪˌləʊz/　美 /'æmɪˌloʊs/	*n.*	直链淀粉
amylopectin	英 /ˌæmɪləʊ'pektɪn/　美 /ˌæməloʊ'pektɪn/	*n.*	支链淀粉
polysaccharide	英 /ˌpɒlɪ'sækəraɪd/　美 /ˌpɑːlɪ'sækəˌraɪd/	*n.*	多糖，多聚糖
lignin	英 /'lɪgnɪn/　美 /'lɪgnɪn/	*n.*	木质素

（续表）

chitosan	英 /ˈkaɪtəʊˌzæn/ 美 /ˈkaɪtəzæn/	n.	壳聚糖
chitin	英 /ˈkaɪtɪn/ 美 /ˈkaɪtɪn/	n.	甲壳素
gelatin	英 /ˈdʒelətɪn/ 美 /ˈdʒelətɪn/	n.	明胶（Fig.1-5）
alginate	英 /ˈældʒɪˌneɪt/ 美 /ˈældʒəˌneɪt/	n.	海藻酸盐
heparin	英 /ˈhepərɪn/ 美 /ˈhepərɪn/	n.	肝素

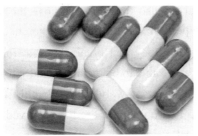

Figure1-5 Capsule shells made from gelatin

例 句

- Cellulose is the oldest and most abundant natural polymer on the earth, is inexhaustible, is mankind's most valuable natural renewable resources.
  纤维素是地球上最为古老、最丰富的天然高分子，是取之不尽用之不竭的，是人类最宝贵的天然可再生资源。

- Starch is a substance that is found in foods such as bread, potatoes, pasta, and rice and gives you energy.
  淀粉是面包、土豆、意大利面和米饭等食物的主要成分，能给你提供能量。

- Trees containing less lignin and more cellulose would both grow faster and also produce more ethanol.
  含低木质素高纤维的树木能够在快速生长的同时又多产乙醇。

- Chitosan, a biologic cationic polyelectrolyte, shows the excellent flocculent capability in water treatment.
  壳聚糖作为生物性的阳离子聚电解质，在水处理方面显示了优异的絮凝性能。

- Chitin is the only basic amylose in nature and has excellent properties of biocompatibility and biodegradability.
  甲壳素是自然界中唯一的碱性多糖，有良好的生物相容性和生物降解性。

- Softgel consists of gelatin, glycerin and water.
  软胶囊由明胶、甘油和水组成。

- At present, one of the biodegradable fibers that cost most study efforts is alginate fiber.
  目前，研究最多的生物可降解纤维之一就是海藻酸纤维。

- Low molecular weight heparin was used for anticoagulation.
  低分子肝素被用于抗凝。

拓展阅读

Polysaccharide

Polysaccharides are long chain polymeric carbohydrates composed of monosaccharide units bound together by glycosidic linkages. Examples include starch, cellulose, chitin, glycogen, and chitosan. Sometimes, polycarbohydrate can be hydrolyzed by using amylase as catalyst, which yields monosaccharide（单糖）and oligosaccharide（寡糖）. Oligosaccharide usually contains 3 to 10 monosaccharide units. Polysaccharides have a general formula of $C_x(H_2O)_y$. When the repeating units in the polymer backbone are six-carbon monosaccharide, also known as pyranose, the general formula can be described as $(C_6H_{10}O_5)_n$. For humans, some polysaccharides, just like starch, can serve as the source of energy, because many organisms in human body can break down starch into glucose. Cellulose cannot be degraded in human organisms, but is able to be metabolized by some protists and bacteria. Cellulose can be obtained through the treatment of wood with ethanol, sodium chlorite, sodium chloride, and finally water. Numerous modified polymeric products derived from cellulose have achieved broad applications in the biomedical field, such as ethylcellulose (EC), cellulose acetate and hydroxypropyl methylcellulose (HPMC). Chitin is a polysaccharide coming from the exoskeleton（[昆]外骨骼）of the arthropod（节肢动物）, which can serve as a good inducer of plant defense mechanisms for controlling diseases.

Unit 19
Product

- 读音：英 /ˈprɒdʌkt/　美 /ˈprɑːdʌkt/
- 释义：*n*. 产物，产品
- 英英释义：A substance formed in a chemical reaction（《柯林斯英汉双解大词典》）.

固定搭配/常用短语：

polymer product	高分子材料制品/聚合物制品（Fig.1-6）
chemical product	化工产品
product quality	产品质量

例 句

- The present invention is suitable for forming large polymer product.
  本发明适用于大型聚合物制品的成型加工。
- The control of molecule weight and molecule weight distribution is often used to obtain and improve certain desired physical properties in a polymer product.
  对聚合物分子量和分子量分布的控制通常用于获得并提升聚合物产品某种期望的物理性质。
- Chitosan is able to react with vanillin aldehyde by Schiff's base reaction in water, a polymer product—VCG can be prepared.
  在水溶液中，壳聚糖可与香草醛发生席夫碱反应，生成聚合物产品 VCG。

Figure1-6　Polymer products in our daily life

常见的高分子制品：

rubber	英 /ˈrʌbə(r)/　美 /ˈrʌbər/	*n.*	橡胶
plastic	英 /ˈplæstɪk/　美 /ˈplæstɪk/	*n.* *adj.*	塑料 塑料的

(续表)

fiber	英 /ˈfaɪbə(r)/ 美 /ˈfaɪbər/	n.	纤维
membrane	英 /ˈmembreɪn/ 美 /ˈmembreɪn/	n.	膜
paint	英 /peɪnt/ 美 /peɪnt/	n.	漆
adhesive	英 /ədˈhiːsɪv/ 美 /ədˈhiːsɪv/	n.	黏合剂
pitch	英 /pɪtʃ/ 美 /pɪtʃ/	n.	沥青
acrylic	英 /əˈkrɪlɪk/ 美 /əˈkrɪlɪk/	adj.	丙烯酸的
nylon	英 /ˈnaɪlɒn/ 美 /ˈnaɪlɑːn/	n.	尼龙
acrylon		n.	腈纶
spandex	英 /ˈspændeks/ 美 /ˈspændeks/	n.	氨纶
rayon	英 /ˈreɪɒn/ 美 /ˈreɪɑːn/	n.	人造丝
hydrogel	英 /ˈhaɪdrə,dʒel/ 美 /ˈhaɪdrədʒel/	n.	水凝胶
liquid	英 /ˈlɪkwɪd/ 美 /ˈlɪkwɪd/	n.	液体
		adj.	液态的
solid	英 /ˈsɒlɪd/ 美 /ˈsɑːlɪd/	n.	固体
		adj.	固态的
gas	英 /gæs/ 美 /gæs/	n.	气体
gaseous	英 /ˈgæsiəs/ 美 /ˈgæsiəs/	adj.	气态的
sticky	英 /ˈstɪki/ 美 /ˈstɪki/	adj.	黏性的
elastic	英 /ɪˈlæstɪk/ 美 /ɪˈlæstɪk/	adj.	有弹性的
inelastic	英 /ˌɪnɪˈlæstɪk/ 美 /ˌɪnɪˈlæstɪk/	adj.	没有弹性的

例 句

- Therefore, fluorosilicone rubber polymer is an ideal candidate for electrolyte membrane.
  因此，氟硅橡胶聚合物是电解质膜的一种较为理想的候选原料。
- Plastic is produced by chemical processes and can be formed into shapes when heated.
  塑料是由化工过程生产的，并且可以通过加热形成特定形状。
- Cotton is a natural fiber, but rayon and nylon are synthetic.
  棉花是天然纤维，但人造丝和尼龙却是合成纤维。

拓展阅读

Rubber

Rubber is one of the most commonly used polymer products in the modern world. From the earliest time, it came to people's life as a natural product, also known as natural rubber. It can be produced from latex, the extract from rubber trees, which mainly consisted by polyisoprene. In the 1820s, rubber products went to market in England. From then on, natural rubber has been extensively used in many products, either alone or in combination with other materials. Benefiting from its large stretch ratio, high resilience, and excellent water-proof ability, natural rubber was used to prepare raincoats, overshoes, gloves, etc., in the 19th century. At the end of 19th century, natural rubber products have not been able to satisfy the industrial demand for rubber-like materials. As a result, synthetic rubber came out in 1909, which was first synthesized by a group in Bayer laboratory. In 1910, Russian chemist Sergey Vasilyevich Lebedev succeeded in making synthetic rubber based on polybutadiene. His book became the bible for the study of synthetic rubber.

From then on, an increasing number of synthetic rubbers have been developed for industrial purposes, e.g. bicycle saddles, car tires, spark plug cables, vibration dampers, seals, and insulators. Especially in World War II, synthetic rubber production has seen an explosive increase. It should be noted that synthetic rubber consisted by carbon-to-carbon backbones is usually susceptible to UV, ozone, heat and other aging factors. To overcome this shortcoming, silicone rubber composed of silicone polymers was synthesized, which offers good resistance to extreme temperatures. However, silicone rubber has been demonstrated to show low tensile strength and poor tear wear properties, which cannot replace the need for classical rubbers.

Unit 20
Data

- 读音：英 /ˈdeɪtə/　美 /ˈdeɪtə/
- 释义：*n.* 数据
- 英英释义：Facts and statistics collected together for reference or analysis　（《新牛津英汉双解大词典》）.

同（近）义词：

information	英 /ˌɪnfəˈmeɪʃn/　美 /ˌɪnfərˈmeɪʃn/	*n.*	信息
statistic	英 /stəˈtɪstɪk/　美 /stəˈtɪstɪk/	*n.*	统计数据
parameter	英 /pəˈræmɪtə(r)/　美 /pəˈræmɪtər/	*n.*	参数
coefficient	英 /ˌkəʊɪˈfɪʃ(ə)nt/　美 /ˌkoʊɪˈfɪʃnt/	*n.*	系数

固定搭配/常用短语：

raw data	原始数据
data processing	数据处理
data analysis	数据分析
experimental data	实验数据
data base	数据库
original data	原始数据

例 句

- Acquiring data such as polymer molecular weight and molecular weight distribution curve of rubber, plastics, and synthetic fibers provide a variety of physical properties of polymer information　（来源：www.nuist.edu.cn）.
  通过获取橡胶、塑料、合成纤维等高分子的平均分子量及分子量分布曲线等数据，可得到高分子的多种物性信息。
- The application cases show that the prediction results match the fact data well, can be used to the effect prediction and long-term planning of polymer flooding.
  实际应用表明，预测结果与实际数据相符程度较高，可以应用于聚合物驱油技术的效果预测和长远规划。
- This paper studies the different data of different polymer and proves the application of boundary

layer treatment by TEM analysis.
本文通过透射电镜分析研究了不同聚合物的不同数据，并验证其在界面处理剂中的应用。

论文中的不同数据类型：

figure	英 /ˈfɪɡə(r)/ 美 /ˈfɪɡjər/	n.	图
table	英 /ˈteɪb(ə)l/ 美 /ˈteɪbl/	n.	表格
scheme	英 /skiːm/ 美 /skiːm/	n.	方案
plot	英 /plɒt/ 美 /plɑːt/	n.	点状图
curve	英 /kɜːv/ 美 /kɜːrv/	n.	曲线
profile	英 /ˈprəʊfaɪl/ 美 /ˈproʊfaɪl/	n.	轮廓图
histogram	英 /ˈhɪstəɡræm/ 美 /ˈhɪstəɡræm/	n.	直方图，柱状图
diagram	英 /ˈdaɪəɡræm/ 美 /ˈdaɪəɡræm/	n.	图表
model	英 /ˈmɒd(ə)l/ 美 /ˈmɑːd(ə)l/	n.	模型
peak	英 /piːk/ 美 /piːk/	n.	峰顶
pattern	英 /ˈpæt(ə)n/ 美 /ˈpætərn/	n.	模式
intensity	英 /ɪnˈtensəti/ 美 /ɪnˈtensəti/	n.	强度
value	英 /ˈvæljuː/ 美 /ˈvæljuː/	n.	价值
rate	英 /reɪt/ 美 /reɪt/	n.	比率，速率
integral	英 /ˈɪntɪɡrəl/ 美 /ˈɪntɪɡrəl/	n.	积分
differential	英 /ˌdɪfəˈrenʃl/ 美 /ˌdɪfəˈrenʃl/	n.	微分
original	英 /əˈrɪdʒən(ə)l/ 美 /əˈrɪdʒən(ə)l/	adj.	原始的

例 句

- Figure 1 illustrates the interaction between a polymer and an application event table.
  图 1 举例说明了聚合物和应用程序事件表之间的相互作用。
- The total synthetic scheme of polymethyl methacrylate was designed with the monomer methyl methacrylate.
  以甲基丙烯酸甲酯为单体，设计出聚甲基丙烯酸甲酯的全合成路线。

拓展阅读

Academic Integrity

Academic integrity is the moral code or ethical policy of academia. It means avoiding plagiarism and cheating, among other misconduct behaviours. Scientific and technological innovation is associated with the destiny of the country. The quality of scientific research in China largely depends on our research environment. "Seeking truth from facts" has always been the basic requirement of China's development. Since the establishment of New China, a vast number of Chinese scientists in China, with noble spiritual realm and good moral sentiments, have made a series of achievements that have attracted worldwide attention. From the older generation of scientists such as Deng Jiaxian, Qian Sanqiang and Guo Yonghuai, to the new generation of role models such as Huang Danian, Nan Rendong and Zhong Yang, this is a group of Chinese scientists with profound knowledge and selflessness. Their scientific achievements can stand the test of time and practice, and their trust-seeking spirit and rigorous attitude show the valuable scientific spirit. Therefore, no matter what level of stress you may find yourself under, you must approach your work with honesty and integrity. Honesty is the foundation of good academic work. Whether you are working on a problem set, lab report, project or paper, avoid engaging in plagiarism, unauthorized collaboration, cheating, or facilitating academic dishonesty. We should pass on the traditions and the culture, and this is a part of our culture.

Unit 21
Symbol

- 读音：英 /ˈsɪmb(ə)l/ 美 /ˈsɪmbl/
- 释义：*n.* 符号，标志
- 英英释义：A symbol is an object that represents or suggests an idea, visual image, belief, action, or material entity, or an item in a calculation or scientific formula presented as a number, letter, or shape.

固定搭配/常用短语：

chemical symbol	化学符号
symbol table	符号表
integral symbol	积分符号

例 句

- What's the chemical symbol for mercury?
  水银的化学符号是什么？
- Precondition research of chemical intelligent exercise is input, distinguish and comprehension various symbol. It is one of technological difficulty about multimedia education software.
  研究化学智能解题的前提是研究各类符号的录入、识别、理解，这是多媒体教育软件中的一个技术难点。

派生词及相关词汇：

formula	英 /ˈfɔːmjələ/ 美 /ˈfɔːrmjələ/	n.	公式
equation	英 /ɪˈkweɪʒn/ 美 /ɪˈkweɪʒn/	n.	等式，方程式
superscript	英 /ˈsuːpəskrɪpt/ 美 /ˈsuːpərskrɪpt/	n.	上标
subscript	英 /ˈsʌbskrɪpt/ 美 /ˈsʌbskrɪpt/	n.	下标
italic	英 /ɪˈtælɪk/ 美 /ɪˈtælɪk/	n.	斜体
variate	英 /ˈveərieɪt/ 美 /ˈverieɪt/	n.	变量
normalize	英 /ˈnɔːməlaɪz/ 美 /ˈnɔːrməlaɪz/	v.	标准化
unknown	英 /ˌʌnˈnəʊn/ 美 /ˌʌnˈnoʊn/	n.	未知量
represent	英 /ˌreprɪˈzent/ 美 /ˌreprɪˈzent/	v.	代表

高分子化学中常用的符号：

符号	名称	中文释义
℃	degree Celsius/ degree centigrade	摄氏度
%	percent	百分号
M_n	number-average molecular weight	数均分子量
M_w	weight-average molecular weight	重均分子量
M_v	viscosity-average molecular weight	黏均分子量
M_p	peak molecular weight	尖峰分子量
$[\eta]$	intrinsic viscosity	特性黏度
mL	milliliter	毫升
L	litre/ liter	（公）升
mg	milligram	毫克
cm	centimeter	厘米
in	inch	英寸
m^2	square meter	平方米
R_h	hydrodynamic radius	流体动力学半径
R_g	mean square radius of gyration	均方回转半径
mol	mole	摩尔
mol/L	mole per liter	摩尔每升

示例：

(1) Liu F, *et al.*, *Polym. Chem.*, 2018, 9: 5024-5031

The diffusion coefficient D and $G(\Gamma)$ can be further converted into the hydrodynamic radius R_h and $f(R_h)$ by using the Stokes-Einstein equation, respectively:

$$D = \frac{k_B T}{6\pi\eta R_h}$$

where k_B, T, and η are the Boltzmann constant, the absolute temperature, and the viscosity of the solvent, respectively.

译：扩散系数 D 和 $G(\Gamma)$ 可以分别用 Stokes-Einstein 方程转化为水动力半径 R_h 和 $f(R_h)$。其中 k_B，T，η 分别为玻尔兹曼常数，溶剂的绝对温度和黏度。

(2) Zhang H, *et al.*, *Macromol. Res.*, 2016, 24: 655-662

M_n (number-average molecular weight), M_w (weight-average molecular weight) and polydispersity values of fluorene-containing polymers were measured via gel permeation chromatography (GPC) (Agilent, USA). The system was operated at a flow rate of 1.0mL/min with DMF as an eluent.

译：含芴聚合物的 M_n (数均分子量)、M_w (重均分子量)和多分散性值是通过凝胶渗透色谱法(GPC)测定的(安捷伦，美国)。以二甲基甲酰胺为洗脱液，系统的流速为 1.0mL/min。

(3) Hu C, *et al.*, *Macromol. Chem. Phys.*, 2018, 219: 1800201

The number-average molecular weight (M_n) and the molecular weight distribution (polydispersity index, M_w/M_n) were measured by gel permeation chromatography (GPC) in THF, where polystyrene was used as standard. The M_n values of P2 and FP2 were determined to be $1.73×10^4$ and $1.75×10^4$ g/mol, respectively, and the M_w/M_n were 1.63 and 1.65.

译：以聚苯乙烯为标准样品，采用凝胶渗透色谱法(GPC)测定了聚合物的数均分子量(M_n)和分子量分布(多分散性指数，M_w/M_n)。测定 P2 和 FP2 的 M_n 值分别为 $1.73×10^4$ 和 $1.75×10^4$ g/mol。多分散性指数分别为 1.63 和 1.65。

Unit 22
Abbreviation

- 读音：英 /əˌbriːviˈeɪʃn/ 美 /əˌbriːviˈeɪʃn/
- 释义：n. 缩写，缩略词
- 英英释义：A shortened form of a written word or phrase used in place of the whole （《韦氏大学英语词典》）。

同根词：

abbreviate	英 /əˈbriːvieɪt/ 美 /əˈbriːvieɪt/	v.	缩写，使省略
abbreviated	英 /əˈbriːvieɪtɪd/ 美 /əˈbriːvieɪtɪd/	adj.	缩写的

abbr. 为 abbreviation 或 abbreviated 的缩写形式。

例 句

- We will use this abbreviation in this paper and throughout the series.
  我们将在本文以及整个系列中使用这个缩略词。
- The abbreviation for polymethyl methacrylate is PMMA.
  聚甲基丙烯酸甲酯的缩写是 PMMA。
- We shall abbreviate "polyvinyl chloride" to PVC in this manuscript.
  在这篇文稿中，我们将把"聚氯乙烯"简称为 PVC。
- Notice we use Notes IDs in abbreviated format.
  注意我们使用缩写形式的 Notes ID。
- This paper deals with the application of inductively coupled plasma-atomic emission spectroscopy abbr. ICP-AES.
  本文讨论了电感耦合等离子体-原子发射光谱法(简称 ICP-AES)的应用。

高分子化学中常用的缩写：

缩写	完整名称	中文释义
PS	polystyrene	聚苯乙烯
PE	polyethylene	聚乙烯
PP	polypropylene	聚丙烯

Unit 22　Abbreviation

（续表）

缩写	完整名称	中文释义
PVA	polyvinyl alcohol	聚乙烯醇
PVC	polyvinyl chloride	聚氯乙烯
ABS	acrylonitrile-butadiene-styrene	丙烯腈-丁二烯-苯乙烯共聚物
PMMA	polymethyl methacrylate	聚甲基丙烯酸甲酯
PVP	polyvinyl pyrrolidone	聚乙烯吡咯烷酮
EVA	ethylene-vinyl acetate copolymer	乙烯-乙酸乙烯酯共聚物

拓展知识

缩写查询网站

通常，在阅读英文文献时，对于相关的缩写，作者会在该词第一次出现时进行说明，或者单独在文后/脚注汇总本文/本页所使用的缩写的全称。但是对于一些专业内公认的缩写，如 FTIR，NMR，CPC 等，在某些文章中也有直接使用的情况。阅读时遇到不理解的缩写，且文中未给出详细解释时，可参考相关缩写查询网站进行查询（Fig.1-7, Fig.1-8）。如下列两个网址：

(1) https://www.abbreviationfinder.org/

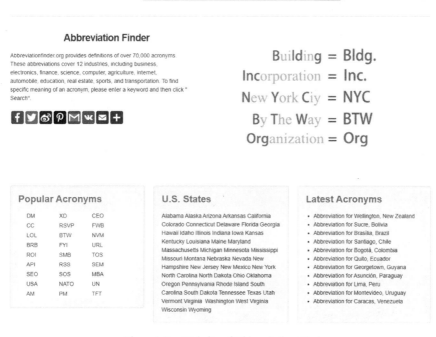

Figure1-7　Website of Abbreviation Finder

(2) https://www.abbreviations.com/

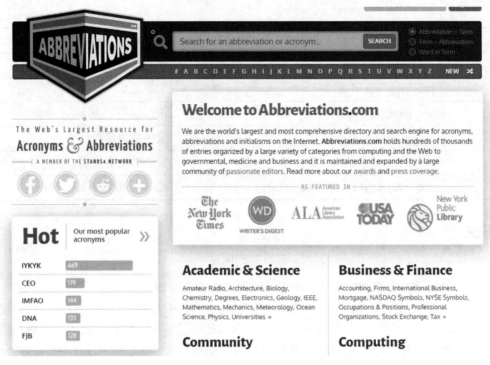

Figure1-8　Website of Abbreviations.com

Unit 23
Configuration

- 读音：英 /kənˌfɪɡəˈreɪʃn/　美 /kənˌfɪɡjəˈreɪʃn/
- 释义：*n.* 构型
- 英英释义：The relative spatial arrangement of atoms in a covalently bonded compound molecule（baike.baidu.com）.

派生词及相关词汇：

isomery	英 /aɪˈsɒməri/　美 /aɪˈsɑːməri/	*n.*	同分异构
isomerization	英 /aɪsɒməraɪˈzeɪʃən/　美 /aɪˌsɒmərəˈzeɪʃən/	*n.*	异构化
chirality	英 /kaɪˈrælɪtɪ/　美 /kaɪˈræləti/	*n.*	手性
chiral	英 /ˈkaɪərəl/　美 /ˈkaɪrəl/	*adj.*	手性的
D-configuration		*n.*	D 构型
L-configuration		*n.*	L 构型
cis	英 /sɪs/　美 /sɪs/	*adj.*	顺式的
trans	英 /trænz/　美 /trænz/	*adj.*	反式的
isotaxy	英 /aɪsəʊˈtæksɪ/　美 /aɪsoʊˈtæksɪ/	*n.*	全同立构
isotactic	英 /ˌaɪsəʊˈtæktɪk/　美 /aɪsoʊˈtæktɪk/	*adj.*	全同立构的
syndiotaxy	英 /sɪndɪəʊˈtæksɪ/　美 /sɪndɪoʊˈtæksɪ/	*n.*	间同立构
syndiotactic	英 /ˌsɪndaɪə(ʊ)ˈtæktɪk/　美 /ˌsɪndaɪoʊˈtæktɪk/	*adj.*	间同立构的
atactic	英 /eɪˈtæktɪk/　美 /eɪˈtæktɪk/	*adj.*	无规立构的
tacticity	英 /tækˈtɪsɪti/　美 /tækˈtɪsɪti/	*n.*	立构规整性

例 句

- The effect of reaction condition on the spatial configuration, molecular weight and distribution of the obtained polymer is investigated theoretically.
  从理论上研究了反应条件对所得聚合物空间构型、分子量和分子量分布的影响。
- Azobenzene and its derivatives are characterized by reversible *cis-trans* isomerization.
  偶氮苯及其衍生物的特点在于可逆顺反异构化。
- This paper introduced the concept of chiral polymer, then reviewed the relationship between chiral

carbon and chiral polymer.

本文介绍了手性聚合物的概念，综述了手性碳与手性聚合物的关系。

- Polyacetaldehyde may exist in atactic, syndiotactic and isotactic forms.
  聚乙醛可以以无规、间同和全同形式存在。

固定搭配/常用短语：

spatial configuration	空间构型
chemical configuration	化学构型
stereo-chemical configuration	立体分子构型
polymer configuration	高分子构型
cis-form	顺式
trans-form	反式
isotactic polymer	全同立构聚合物（Fig.1-9）
syndiotactic polymer	间同立构聚合物
atactic polymer	无规立构聚合物

Figure1-9 Illustration of polypropylene with different configurations

拓展阅读

Cis-1,4-Polyisoprene and *Trans*-1,4-Polyisoprene

Polyisoprene is a collective name for polymers that contains different isomers: *cis*-1,4-polyisoprene, *trans*-1,4-polyisoprene, 1,2-polyisoprene, and 3,4-polyisoprene. Under the catalysis of Ziegler-Natta catalyst $TiCl_4/Al(i\text{-}C_4H_9)_3$, the highly purified product *cis*-1,4-polyisoprene can be achieved following a coordinative chain polymerization. *Cis*-1,4-polyisoprene is an amorphous elastomer at room temperature, which is a component of natural rubber and can also be collected from the sap of rubber trees. *Cis*-1,4-polyisoprene is primarily used for tires, which can also be used for preparing condoms, belting, raincoats, and footwear. When $VCl_3/Al(i\text{-}C_4H_9)_3$ is used as catalyst, *trans*-dominant polyisoprene is formed. *Trans*-1,4-polyisoprene is a semi-crystalline polymer at room temperature, which is much harder than *cis*-1,4-polyisoprene. *Trans*-1,4-polyisoprene has been widely used as electrical insulators and as component of golf ball (Fig.1-10). Other isomers, including 1,2-polyisoprene and 3,4-polyisoprene, have not been widely used like *cis*-1,4-polyisoprene and *trans*-1,4-polyisoprene.

Unit 23　Configuration

Figure 1-10　Structure of *cis*-1,4-polyisoprene and *trans*-1,4-polyisoprene and their application in materials

Unit 24
Conformation

- 读音：英 /ˌkɒnfɔːˈmeɪʃn/ 美 /ˌkɑːnfɔːrˈmeɪʃn/
- 释义：*n.* 构象
- 英英释义：A spatial arrangement of the atoms of a molecule, especially one that results from rotation around a single bond (Microsoft Bing).

派生词及相关词汇：

rotation	英 /rəʊˈteɪʃn/ 美 /roʊˈteɪʃn/	n.	旋转
random	英 /ˈrændəm/ 美 /ˈrændəm/	adj.	无规的
coil	英 /kɔɪl/ 美 /kɔɪl/	n.	线团
barrier	英 /ˈbæriə(r)/ 美 /ˈbæriər/	n.	位垒
segmer	英 /ˈsegmə/ 美 /ˈsegmə/	n.	链段
helical	英 /ˈhelɪkl/ 美 /ˈhelɪkl/	adj.	螺旋形的
zigzag	英 /ˈzɪgzæg/ 美 /ˈzɪgzæg/	n. adj.	锯齿形 锯齿形的（Fig.1-11）

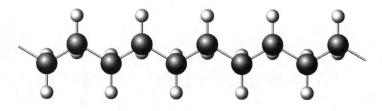

Figure1-11　Zigzag conformation of polyethylene

固定搭配/常用短语：

potential energy	势能，位能
gauche conformation	旁氏构象
trans conformation	反式构象
helical conformation	螺旋构象
zigzag conformation	锯齿构象
random coil	无规线团

（续表）

end-to-end distance	末端距
mean square end-to-end distance	均方末端距
root-mean-square end-to-end distance	均方根末端距
unperturbed chain	无扰链
freely jointed chain	自由连接链
freely rotating chain	自由旋转链

例 句

- The relationship between conformation and configuration of polypropylene and their rotatory direction is analysed by applying the rule of helix theory proposed by author.
  利用作者提出的螺旋理论，具体分析了聚丙烯的构象、构型和其旋光方向之间的关系。
- Flory proposed the random coil model of amorphous polymer in 1960s.
  弗洛里在20世纪60年代提出了非晶态聚合物的无规线团模型。
- On the other hand, the stability of helical conformation of polypropylene decreased with the increasing temperature.
  另一方面，聚丙烯螺旋构象的稳定性随着温度的升高而降低。
- The concept of conformation is different from configuration, the former of which can be changed by the rotation of C—C bonds.
  构象的概念不同于构型，前者可以通过C—C键的旋转而改变。
- The model of freely jointed chain only assumes a polymer as a random walk and overlooks any kind of interactions among monomers.
  自由连接链模型假设聚合物为随机游走，不考虑单体间的任何相互作用。

拓展阅读

Polypropylene and Its Conformation

Polypropylene is a commonly used thermoplastic polymer with a wide variety of applications, which can be synthesized via chain growth polymerization from propylene. Polypropylene can be categorized as atactic polypropylene (APP), syndiotactic polypropylene (SPP), and isotactic polypropylene (IPP) depending on the difference in configuration. IPP is a well known semi-crystalline polymer, which has been demonstrated to exhibit a typical polymorphic behavior. Molecular chains of IPP adopt a helical conformation and can be organized into several spatial arrangements resulting α, β, γ crystal forms or mesophase, which are relying on the crystallization conditions. In a helical conformation, a sequence of repeat units in a polymer is twisted into a helix. Helical conformation is also the most common structural arrangement in the secondary structure of proteins. As compared to configuration, the conformation of polymers is more difficult to be measured. In the 1980s, Noda proposed generalized two dimensional IR (2D-IR) correlation spectroscopy. Since then 2D-IR spectroscopy gets wide applications in various research fields. Nowadays, more and more scholars in polymer science use this technique to analyze polymers' conformation.

Unit 25
Crystal

- 读音：英 /'krɪst(ə)l/ 美 /'krɪst(ə)l/
- 释义：*n.* 结晶
- 英英释义：A crystal is a small piece of a substance that has formed naturally into a regular symmetrical shape（《柯林斯英汉双解大词典》）.

Figure1-12 · Image of sphaerocrystal

派生词及相关词汇：

crystalline	英 /'krɪstəlaɪn/ 美 /'krɪstəlaɪn/	*adj.*	水晶制的
crystallization	英 /ˌkrɪstəlaɪ'zeɪʃ(ə)n/ 美 /ˌkrɪstələ'zeɪʃ(ə)n/	*n.*	结晶化
amorphous	英 /ə'mɔːfəs/ 美 /ə'mɔːrfəs/	*adj.*	无定形的，非晶形的
semicrystalline	美 /'semi 'krɪstəlaɪn/	*adj.*	半晶状的
syngony	英 /'sɪngənɪ/ 美 /'sɪngənɪ/	*n.*	晶系
monocrystal	英 /'mɒnəkrɪstəl/ 美 /ˌmɑːnə'krɪstl/	*n.*	单晶
polycrystal	英 /ˌpɒlɪ'krɪst(ə)l/ 美 /ˌpɒli'krɪstl/	*n.*	多晶
sphaerocrystal	英 /'sfɪərəʊˌkrɪstəl/ 美 /'sfɪroʊˌkrɪstəl/	*n.*	球晶（Fig.1-12）
shish-kebab	英 /'ʃɪʃ kɪbæb/ 美 /'ʃɪʃ kɪbɑːb/	*n.*	串晶
dendrite	英 /'dendraɪt/ 美 /'dendraɪt/	*n.*	树枝状晶体
lamellae	英 /lə'meliː/ 美 /lə'meli/	*n.*	片晶
orientation	英 /ˌɔːriən'teɪʃ(ə)n/ 美 /ˌɔːriən'teɪʃn/	*n.*	取向
crystallinity	英 /ˌkrɪstə'lɪnətɪ/ 美 /ˌkrɪstə'lɪnətɪ/	*n.*	结晶度

Unit 25　Crystal

固定搭配/常用短语：

crystalline region	晶区
amorphous region	非晶区
degree of crystallinity	结晶度
crystal cell	晶胞
monoclinic syngony	单斜晶系
fibrous crystal	纤维晶
fringed-micelle model	缨状微束模型
folded chain model	折叠链模型
switchboard model	插线板模型
liquid crystal	液晶
crystallization temperature	结晶温度

例 句

- The crystallization and orientation behaviors of polymer film were discussed in this study. The orientation features in crystalline and amorphous regions were compared.
  本研究讨论了聚合物薄膜的结晶行为和取向行为，分析了晶区和非晶区取向的不同特点。
- The crystallinity of polyethylene decreased with increase of content of short branched chain methyl.
  聚乙烯随着短支链甲基含量的增加，结晶度降低。
- Liquid crystals are considered to be intermediate between liquid and solid（来源：牛津英文词典）.
  液晶被认为是介于液态和固态之间的中间体。
- Researchers proposed various models to explain the crystallization behaviors in polymers, including fringed-micelle model, folded chain model and switchboard model.
  研究人员提出了多种模型来解释聚合物的结晶行为，包括缨状微束模型、折叠链模型和插线板模型。

拓展阅读

Crystallization of Polymers

　　Polymers can crystallize upon cooling from melting, mechanical stretching or solvent evaporation. However, fully crystalline polymers are impossible to be obtained, which is different from small molecules. Some chains in polymers can fold together to form ordered regions upon cooling from melting, mechanical stretching or solvent evaporation. The fraction of the ordered moieties in polymer is characterized by the degree of crystallinity, which typically ranges between 10% and 80%. Crystallized polymers can be named "semicrystalline". According to the literature, density measurement, differential scanning calorimetry (DSC), X-ray diffraction (XRD), infrared spectroscopy and nuclear magnetic resonance (NMR) are be employed to measure the degree of crystallinity. Sometimes, crystal growth in polymers can be observed by a polarizing microscope. Crystal growth can occur for temperatures below the melting temperature (T_m) and above the glass transition temperature (T_g). Higher temperatures would destroy the molecular arrangement. When the temperature is below the T_g, the movement of molecular chains is frozen. Crystallization of polymers, crystal growth, as well as the degree of crystallinity, are very important in polymer physics, which can affect the mechanical, optical, thermal and chemical properties of the polymer.

拓展篇

Unit 26
Introduce Yourself

　　自我介绍是我们来到新环境、遇到新朋友、接触新事物的一个必不可少的环节。现代生活中，英文自我介绍的使用频率逐渐增多，如外企面试、考研复试、出国交流、签证面签等。所以，英文自我介绍的水平直接关系到你给别人的第一印象的好坏、交往的顺利与否，甚至考研和工作能否成功上岸。如果在英文自我介绍中能够展示出自己的专业外语水平，将显著提高面试官和听众对你的印象分。因此，准备一份专业外语版的自我介绍，是工作和学习中至关重要的环节。

　　英文自我介绍的设计应兼顾受众的水平，如仅仅是新同学或者新朋友的初次见面，或课堂中的自我介绍，应主要关注家庭（family）、家乡（hometown）和爱好（habit）；而如果是考研复试等专业性较强的环节，则应以所学专业（major）、课程（course）、研究兴趣（research interests）等为主要介绍内容。本节我们将对以上关键要素和表达方式进行总结，并根据不同的需求，设计不同类型的自我介绍。

要素 1：Family

例 句

- In past decades, China had a single-child policy. So I grew up in a warm family of three.
  在过去的几十年里，中国一直实行独生子女政策。所以我在一个温暖的三口之家长大。
- Our family is very pleased to have a such a good family tradition and honor the elderly （来源：有道词典）.
  很高兴我们家有一个这么好的孝敬老人的家风。
- In my opinion, family is a source of emotional support, warmth and comfort.
  在我看来，家庭是情感支持、温暖和安慰的来源。

　　Family 部分的介绍要注意避免流于平庸。传统的家庭介绍往往局限于介绍有几位家庭成员，分别是谁。本节中给出的例句提供了三种不同的思路，其中涉及 single-child（独生子女）、family tradition（家风）等具有中国特色的描述形式，以及对于家庭的观点。例如，可以将家庭比喻为心灵的港湾、充电站等形象化的事物，也可以描述家庭在心目中的重要地位等。

要素 2：Hometown

例 句

- A must-eat in Baoding city is the donkey burger which attracts both local and foreign visitors with its old history and fantastic taste.

Unit 26 Introduce Yourself

保定有一个必吃的美食就是驴肉火烧，它凭借悠久的历史和独特的美味吸引了国内外游客。

- It is located in Northern China. It lies on the eastern side of the Songhua River.
  它位于中国北方，松花江的东岸。
- Baoding is praised as the "champion city" as more than hundred world champions were born in this city.
  保定被誉为"冠军之城"，诞生了上百位世界冠军。
- The history of Beijing can be traced back to thousands of years when the king of Yan in Zhou dynasty built his capital city here.
  北京的历史可以追溯到几千年前，周朝的燕王在此创建了他们的首都。

　　Hometown 部分的介绍要注意突出特色，避免使用一些泛泛的形容词，如用 beautiful, famous, attractive 等形容自己的家乡，难以给人留下深刻印象。上述例句提供了描绘地区特色的几种方式，例如可使用地方美食、方位、别称等表现城市的特色。

要素 3：Leisure activity (Habit)

例 句

- In leisure time, I prefer to hang out with my girlfriend.
  在闲暇时光，我喜欢和我的女朋友去闲逛。
- Attending academic conference is a primary way of socializing with others and improving academic levels.
  参加学术会议是与他人交往和提高学术水平的一种主要方式。
- Playing basketball can help us burn up calories and lose weight.
  打篮球可以帮助我们燃烧卡路里和减肥。

要素 4：Major

高分子相关专业的准确翻译：

高分子材料与工程	polymer material and engineering
高分子化学与物理	macromolecular chemistry and physics

例 句

- I major in polymer material and engineering / macromolecular chemistry and physics. Polymer is a natural or synthesized substance consisting of compounds with large molecular weight. It is very familiar in our daily life. The plastic bottles, our clothes, tires and packing bags are mainly made of polymers. The objective of my study is to develop novel kinds of polymers with better properties, which can meet a broader application.
  我的专业是高分子材料与工程/高分子化学与物理。聚合物是由大分子量化合物组成的天然或合成的物质。它在我们的日常生活中是非常常见的。塑料瓶、我们的衣服、轮胎和包装袋都是由聚合物制成的。我的研究目标是开发性能更好的新型聚合物，满足更广泛的应用。

要素 5：Course

高分子相关课程的准确翻译：

高分子化学	polymer chemistry
高分子物理	polymer physics
高分子加工	polymer molding
医用高分子	biomedical polymer
聚合物表征	polymer characterization
聚合物制备工程	polymer preparation engineering
功能高分子	functional polymer
涂料和黏合剂	coatings and adhesives

例 句

- Moreover, I have completed the senior course of polymer material and engineering, including polymer chemistry, polymer physics, polymer molding, functional polymer and polymer characterization. I find that polymer science is so interesting and exciting. I'm hoping to get onto a PhD program upon polymer science.
  而且，我已经完成了高分子材料与工程的高级课程，包括高分子化学、高分子物理、功能高分子、高分子加工和聚合表征。我发现高分子科学太有趣了，太令人兴奋了。我希望继续攻读高分子科学的博士学位。
- This not only led to my 4.03 GPA in my specialty but also ranked first in my class.
  这不仅使我在我的专业获得了 4.03 的绩点，并且获得了班级第一的排名。
- Radical chain polymerization is an important content to study in the course of "polymer chemistry"（来源：有道词典）.
  自由基聚合是《高分子化学》学习的重要内容。
- I went through the entire program of polymer physics in the third year, and I loved it so much.
  我在第三年完成了高分子物理课程的全部内容，我太喜欢这门课程了。
- Flory's theory serves as an excellent guide to a vast subject in polymer physics, and will suggest plenty of further lines of inquiry.
  弗洛里的理论为聚合物物理学这一庞大学科提供了很好的指导，并将提出许多进一步的研究方向。

要素 6：Research interests

Research interests 主要指研究兴趣，自我介绍时的研究兴趣应参照高分子领域常见研究方向进行选择，下表中给出了目前高分子领域的常见研究方向，统计来源于 2019 年全国高分子学术论文报告会。

Unit 26 Introduce Yourself

高分子化学	polymer chemistry
生物大分子	biomacromolecule
高分子物理	polymer physics
高分子物理化学	physical chemistry in polymer science
高分子理论	polymer theory
生物医用高分子	biomedical polymer
仿生与智能高分子	biomimetic and intelligent polymers
光电高分子	optoelectronic polymer
能源高分子	energy polymer
超分子体系	supramolecular system
高性能高分子	high performance polymer
高分子成型加工	polymer molding
高分子共混与复合体系	polymer blend and composite system
片型高分子	two-dimensional polymer
生物基高分子	bio-based polymer

例句

- I know you are very interested in the study of polymer theory and always attach importance to the broad discipline of thoughts and theoretical study in polymer chemistry, polymer physics and polymer molding.
我知道你们对聚合物理论的研究非常感兴趣，并且一直重视高分子化学、高分子物理和高分子成型的广泛的学科思想和理论研究。
- Biomedical polymer has become one of the hottest research areas in polymer science. I am also interested in it!
生物医用聚合物已成为高分子科学研究的热点之一。对此我也很感兴趣！
- Some areas of in polymer theory have few prospects of a commercial return, but they can promote scientific advances.
聚合物理论中的一些领域几乎没有商业回报的前景，但它们可以促进科学进步。

要素 7：Education

Education 主要指教育经历，下表中总结了与教育相关的词汇：

幼儿园	kindergarten
小学	primary school elementary school
初中	middle school
高中	senior school
大学入学考试（高考）	college entrance examination
大学	university
学士	bachelor

（续表）

硕士	master
博士	doctor philosophy doctor, PhD
博士后	postdoctor
访问学者	visiting scholar

例 句

- I passed Chinese College Entrance Examination and got into Tsinghua University in 2020.
  我在高考中取得了好成绩，于 2020 年进入清华大学学习。
- My senior and middle school does not be located in my hometown, I can take care of myself, and adapt to the new environment easily.
  由于我的初中和高中不是在我家乡读的，我可以照顾好自己，能很容易适应新的环境。
- I got the bachelor of science at College of Chemistry and Environmental Science, Hebei University and stayed there to take my PhD in macromolecular chemistry and physics.
  我在河北大学化学与环境科学学院获得了理学学士学位，并在那里攻读高分子化学与物理博士学位。

以上为英文自我介绍的七种主要要素。除此之外，根据不同场合，仍有一些要素需要注意，例如工作（job）、留学经历（oversea study experience）、健康状况（physical condition）、技能（skill）等。可根据具体需要，与前述七种主要要素搭配设计。英文自我介绍的场景一般为口语表达，限定因素也以时间为主，因此应将语速考虑在内。英语表达语速因人而异，与个人习惯以及英语水平直接相关。以播音员语速为例，英国广播公司（British Broadcasting Corporation）的正常语速为 150～180 词/分钟，美国之音（Voice of America, VOA）的正常语速为 130～160 词/分钟，慢速英语语速为 90～100 词/分钟。因此，一份 1～2 分钟的英文自我介绍应将总词数控制在 100～300 词。

范文 1：

My name is Li Ming. In past decades, China had a single-child policy. So I grew up in a warm family of three. My mother is a doctor. She has been busying in the drug clinical trial, especially after the break out of COVID-19. My father is a government employee. He told me that our family tradition was to be honest and upright. In my opinion, family is a source of emotional support, warmth and comfort. My hometown is Baoding. Baoding is praised as the "champion city" as more than hundred world champions were born in this city. A must-eat in Baoding city is the donkey burger which attracts both local and foreign visitors with its old history and fantastic taste. Because it is very near to and located on the South of Beijing, Baoding city is also praised as the southern gate of capital. I am a big fan of football. Playing football can help us burn up calories and lose weight. Moreover, playing football is a primary way of socializing with others and improving communications. I hope to meet some new friends here. Thanks for your attention.

译文：我叫李明。在过去几十年里，中国实行独生子女政策。所以我在一个温暖的三口之家长大。我妈妈是医生。她一直忙于药物临床试验工作，尤其是在 COVID-19 爆发后。我爸爸是政府工作人员。他告诉我，我们家的家风是诚实和正直。在我看来，家庭是情感支持、温

Unit 26 Introduce Yourself

暖和舒适的来源。我的家乡是保定。保定被誉为"冠军城市",因为有 100 多名世界冠军出生在这个城市。保定市的必吃食品是驴肉火烧,它以其古老的历史和美妙的味道吸引着国内外的游客。由于保定市离北京很近,位于北京以南,所以也被誉为首都的南大门。我是足球迷,踢足球可以帮助我燃烧卡路里和减肥。此外,踢足球还是与他人社交和改善沟通的良好方式。我希望能在这里结识一些新朋友。谢谢你的关注。

建议:上述英文自我介绍主要关注家庭(family)、家乡(hometown)和爱好(habit)三种要素,表达也以通俗的口语化表达为主,因此较为适用于课堂、聚会、入学等非正式场合。

范文 2:

I am Zhang San. I am currently a senior student studying polymer material and engineering with a minor in chemistry at College of Chemistry and Environmental Science, Hebei University. Polymer is a natural or synthesized substance consisting of compounds with large molecular weight. It is very familiar in our daily life. The plastic bottles, our clothes, tires and packing bags are mainly made of polymers. The objective of my study is to develop novel kinds of polymers with better properties, which can meet a broader application. I have completed the major courses of polymer material and engineering, including polymer chemistry, polymer physics, polymer molding, functional polymer and polymer characterization. My hard work not only led to my 4.03 GPA in my specialty but also made me rank first in my class. Now, I am very interested in polymer theory. Some areas of polymer theory have few prospects of a commercial return, but they can promote scientific advances. It sounds great.

译文:我是张三。我目前是河北大学化学与环境科学学院的大四学生,主修高分子材料与工程,辅修化学。聚合物是一种天然或合成的物质,由大分子量的化合物组成。它在我们的日常生活中很常见。塑料瓶、我们的衣服、轮胎和包装袋都是由聚合物制成的。现阶段我学习的目的是开发性能更好的新型聚合物,以满足更广泛的应用。我已经完成了高分子材料与工程的主要课程,包括高分子化学、高分子物理、高分子成型、功能高分子和高分子表征。我的努力学习不仅让我的专业平均绩点达到了 4.03,而且在班级排名第一。现在,我对聚合物理论非常感兴趣。聚合物理论中的一些领域几乎没有商业回报的前景,但它们可以促进科学进步。这听起来好棒。

建议:范文 2 中的英文自我介绍主要关注专业(major)、课程(course)、研究兴趣(research interests)和教育(education)几种要素。其表达较范文 1 中更为书面化,且表达中涉及较多专业词汇,较适用于研究生复试面试等正式场合。

Unit 27
Write a Letter

书信（letter）是自古以来人类传递信息的一种主要媒介，是中西方文化沟通交流中的重要形式，距今已有几千年的历史。20 世纪以来，随着互联网科学技术的飞速发展，电子邮件（E-mail）逐步取代了传统纸质版的书信，成为了新时代人类沟通的主要形式。虽然书信的载体由纸质向电子化逐渐发展，但是以电子邮件为载体的这种书信沟通在当代仍具有十分重要的作用。在专业领域，书信形式具有很多种，用途也不尽相同，下表中汇总了专业领域内不同类别书信的主要形式和用途。

Letter of application	申请信 主要用于申请奖学金、求职等方面
Letter of invitation	邀请信 主要用于邀请专业领域内学者来校访问、出席会议等方面的正式书面邀请；也可以用作书面邀请投稿的约稿或者邀稿
Letter of recommendation	推荐信 主要用于推荐某人去报名考试、读研、进修以及工作等
Cover letter	投稿信 主要用于投稿时与编辑的沟通使用，一般用于初次投稿介绍文章的主要创新点
Response letter	回复信 主要用于文章返回修改后与编辑和审稿人的沟通，一般包含对文章主要修改部分的说明，以及对审稿人提问的回复

对于上述英文信件，一般均包括目的（objective）、信息（information）、收尾（ending）、祝福（complimentary）、签字落款（signature）等要素。

要素 1：Objective

例 句

- I'm writing this letter to apply for admission into Harvard university to pursue my PhD degree (*letter of application*).
  我写这封信是想申请进入哈佛大学攻读博士学位（申请信）。
- I am writing to express my appreciation to you for your support and great kindness (*letter of thanks*).
  我写这封信是为了感谢您的支持和好意（感谢信）。

- On behalf of the American Chemical Society, I am delighted to cordially invite you to join the 2021 ACS Spring Meeting, which will be held in Columbia University in the City of New York, on June 13-15, 2022 as a speaker (*letter of invitation*).
  我很高兴代表美国化学学会诚挚地邀请您参加 2021 年 ACS 春季会议，该会议将于 2022 年 6 月 13-15 日在纽约市哥伦比亚大学举行（邀请信）。

- As an associate professor of College of Chemistry and Environmental Science in Hebei University, I would like to take the opportunity to offer a formal recommendation for Mr Zhang San for the master program of your university (*letter of recommendation*).
  作为河北大学化学与环境科学学院的一名副教授，我想借此机会正式推荐张三同学攻读贵校的硕士研究生（推荐信）。

- Enclosed please find the manuscript entitled "Synthesis of glucose-based hyperbranched glycopolymers" submitted to Macromolecules as a research article coauthored by Li Si, Zhang San and Wang Wu (*cover letter*).
  随函附上由李四、张三和王五合著的题为"葡萄糖基超支化糖聚合物的合成"的手稿，作为研究文章提交给 Macromolecules 期刊，请查收（投稿信）。

  我们写书信的时候经常以"How do you do?"或者"How are you?"的口语式表达开头，这是不恰当的。书信的开头，在称谓之后，应以目的句直接开头，简洁明了。

要素 2：Information

在第一句说明信件的目的后，应根据目的对需要说明的信息进行描述。例如在申请信中应主要介绍自己的个人情况，此处可参考"自我介绍"课程中的相关描述；申请信应将邀请细节，如时间、行程、其他参与人以及劳务费等情况进行说明；推荐信中一般会对所推荐学生的学习工作情况进行简要介绍；求职信和回复信则应针对文章的创新性和修改信息进行描述。

例 句

- Having known Zhang San for more than four years, I have no doubt in his ability to undertake a demanding master course like yours. Furthermore, his in-depth mastery of professional knowledge, familiarity with various experimental operations, and good communication skills have won unanimous recognition from other research group members and me (*letter of recommendation*).
  认识张三已经四年多了，我毫不怀疑他有能力攻读像你们这样要求很高的硕士课程。此外，他对专业知识的深入掌握，对各种实验操作的熟悉，以及良好的沟通技巧，赢得了研究小组其他成员和我的一致认可（推荐信）。

- The meeting will be held on October 10-12, 2022 at Dubai, UAE and aimed to expand its coverage in the areas of polymer science where expert talks, young researcher's presentations will be inspired and keep up your enthusiasm (*letter of invitation*).
  本次会议将于 2022 年 10 月 10 日至 12 日在阿联酋迪拜举行，旨在扩大其在聚合物科学领域的覆盖范围，届时将举行专家讲座、年轻研究员的演讲，会议将激发您的灵感并保持您的热情（邀请信）。

- The work has not been published before, and the publication has been approved by all coauthors and the responsible authorities. We find the results are of great interests, and we decide to write this article to share our experimental studies on the famous journal (*cover letter*).
这篇文章之前从未发表过，并且这篇文章已获得所有合著者和主管部门的认可批准。我们发现这些研究结果非常有趣，在这本著名杂志上我们决定发表这篇文章来分享我们的实验研究（投稿信）。

要素 3：Ending

例 句

- If you have any further questions, please feel free to contact me at your convenient time.
如果您还有任何问题，请在您方便的时候与我联系。
- I am looking forward to hearing from you in the earliest possible time.
我期待着能尽快收到您的来信。

要素 4：Complimentary

书信的主体内容结束后，一般在结尾处顶格写上签名前的谦语或祝福词，通常为

- (Yours) Sincerely
- (Yours) Faithfully
- (Yours) Affectionately
- With best wishes
- Best regards

要素 5：Signature

书信末尾的落款处，除包含作者的名字或电子签名等，一般还应包括以下信息：

- Title: Dr. / Prof./ Ms. /Mr./…
- Position: Dean / Chairman / President / Director of Department /…
- Company: University / College / Laboratory/…
- No./street: e.g. No.180 WuRoad/…
- City: Baoding / Beijing / Shanghai / New York/ Zurich / Brussels / …
- Zipcode: 100000/…
- Country name

此外，在英文信件中，人们出于简洁明了的目的，经常对一些常见的词汇进行缩写，下表中对常见的缩写进行了汇总。

carbon copy	CC	抄送
care of	C/O	由……转交
per procurationem	PP	由……代表

		（续表）
as soon as possible	ASAP	尽快
signed	sgd	签字
enclosure	Encl	附件

范文 1：**Letter of application**

Dear Sir/Madam,

  I am writing to apply for a British Council scholarship to study in Britain. I did my four years' basic learning in polymer material and engineering at Hebei University. After this I successfully completed a one-year language course at Beijing Foreign Studies University and passed International English Language Testing System (IELTS) with a score at 7.5.

  As a result of the publication in 2020 of my article in "*Journal Name*" entitled "Synthesis and Characterization of Ladder-type Polymers", which was based on an experiment over 6 months. I made contact with Dr. Zhang San and Prof. Li Si in Hebei University. I have received their help and advice with my current research project.

  Despite their help, I feel that I cannot make real progress in this field unless I can actually have day-to-day contact with these specialists. I would therefore like to continue my research at either the University or the Research Laboratory as soon as my probationary period here is completed. I prefer to study in Britain for two years.

  I enclose a detailed curriculum vitae and an offprint of my article, together with two letters of recommendation from Dr. Zhang San and Prof. Li Si.

Yours faithfully,
Wang Wu

译文
尊敬的先生/女士：
  我写信是为了申请英国文化协会奖学金来英国学习。我在河北大学学习了四年的高分子材料与工程基础知识。在此之后，我成功地在北京外国语大学完成了为期一年的语言课程，并以 7.5 分通过了雅思考试。
  2020 年，我在《（期刊名）》杂志上发表了一篇题为"梯形聚合物的合成与表征"的文章，这篇文章是基于一项为期 6 个月的实验。我与河北大学的张三博士和李四教授取得了联系。我在目前的研究项目中得到了他们的帮助和建议。
  尽管有他们的帮助，我仍觉得不能在这个领域取得真正的进展，除非我能与这个领域的更多专家进行日常接触。因此，我想在这里的试用期一结束，就到英国的大学或研究实验室继续学习研究两年。
  随函附上我的详细简历和文章的副刊，以及张三博士和李四教授的两封推荐信。

您诚挚的，
王五

范文 2：Letter of application

Dear Sir or Madam,

  As an associate professor of College of Chemistry and Environmental Science in Hebei University, I would like to take the opportunity to offer a formal recommendation for Mr Guo Ziyang for the master programme of your university. Having known Ziyang for more than four years, I have no doubt in his ability to undertake a demanding master course like yours.

  I think the biggest advantage of Ziyang is his great potential in scientific research. He joined my research group in his freshman year. In the group, he completed all the tasks assigned to him well and took the initiative to grasp research skills. For example, in the paper named *Facile Grafting of Ionic Liquids onto Halloysite Nanotubes via An Atom Transfer radical Polymerization Method*, he did a good job of the product characterization that he was responsible for. The paper has been accepted and published in *Journal of Polymer Materials*. Furthermore, his in-depth mastery of professional knowledge, familiarity with various experimental operations, and good communication skills have won unanimous recognition from other research group members and me.

  To my knowledge, Ziyang's academic performance is among the best as well. In his senior year, he took my course of Professional English for Polymer Materials and ranked the 3rd with a high score of 90 points. In addition, he has participated in the college students' innovation and entrepreneurship training program held by the university four times. Among them, two projects initiated by him as the leader have been rated as the university-level project, which is a recognition of his innovative spirit in research and proof of his leadership skills.

  In all, I am confident that Ziyang has more than enough abilities to successfully complete your Master's degree course. Should you have any questions or concern about what I have provided, please do not hesitate to contact me directly.

Yours sincerely,

Zhang Macro
Associate Professor
Hebei University
Tel: 86 139××××
Email: ××××@hbu.edu.cn
Add: No. 180, Wusi Dong Road, Baoding City, Hebei Province, China

译文
尊敬的先生/女士：
  作为河北大学化学与环境科学学院的副教授，我想借此机会正式推荐郭子洋同学攻读贵校的硕士学位。认识子洋已经四年多了，我对他的能力充满信心，我认为他有能力攻读像你们这样要求很高的硕士课程。
  我认为子洋最大的优势是他在科学研究方面的巨大潜力。他在大一时加入了我的研究小组。在小组中，他很好地完成了分配给他的所有任务，并主动掌握了研究技能。例如，在题为

"通过原子转移自由基聚合法将离子液体轻松接枝到埃洛石纳米管上"的论文中,他出色地完成了他负责的产物表征方面的研究工作。该论文已被《高分子材料》杂志接受并发表。此外,他对专业知识的深入掌握,对各种实验操作的熟悉,以及良好的沟通技巧,赢得了包括我在内的研究小组成员的一致认可。

据我所知,子洋的学习成绩也是很好的。他在大四时选修了我的"高分子材料专业英语"课程,以 90 分的高分排名第三。此外,他还四次参加学校举办的大学生创新创业培训项目。其中,他作为领导者发起的两个项目被评为大学级项目,这是对他研究创新精神的认可,也是对他领导能力的认可。

综上,我相信子洋有足够的能力可以成功完成你们的硕士学位课程。如果您对我提供的信息有任何疑问或担忧,请直接与我联系。

张麦克
河北大学副教授
电话:86 139××××
邮箱:××××@hbu.edu.cn
地址:中国河北省保定市五四东路 180 号

范文 3:Cover letter

Dear editor,

Enclosed please find the manuscript submitted to *Chemical Communications* as a communication that is entitled "A Facile One-step Grafting of Polyphosphonium onto Halloysite Nanotubes Initiated by Ce(Ⅳ)" coauthored by Author-1, Author-2, Author-3, Author-4 and Author-5.

The work has not been published before, and the publication has been approved by all coauthors and the responsible authorities at our colleges where the work has been carried out.

In the manuscript, we proposed a facile strategy to graft polyphosphonium onto halloysite nanotubes (HNTs) in one step by constructing a redox-initiating system consisted of the hydroxyl groups on HNTs together with the supernormal valence transition-metal Ce(Ⅳ). The polyphosphonium-grafted were immersed into sodium alginate solution to achieve a uniform hydrogel. Interestingly, the prepared hydrogel shows a desirable antibacterial activity. We find the results are of great interests, and we decide to write this article to share our experimental studies on the famous journal—*Chemical Communications*.

It will be greatly appreciated, if we get your kind help and support.

With best wishes.
Sincerely yours

Zhang Macro
Associate Professor
Hebei University
Tel: 86 139××××

Email: ××××@hbu.edu.cn
Add: No. 180, Wusi Dong Road, Baoding City, Hebei Province, China

译文
尊敬的编辑：

　　随函附上由五位作者合著的题为"四价铈引发的埃洛石纳米管一步接枝磷基聚合物"的投给贵刊《化学通讯》的通讯稿，请查收。

　　这篇文章之前从未发表过，并且该文已获得所有合著者和开展该工作的学院的主管单位的认可批准。

　　本项研究中，我们提出了一种简单的方法，通过构建由埃洛石纳米管上的羟基和超常价过渡金属四价铈组成的氧化还原引发体系，一步将磷基聚合物接枝到埃洛石纳米管上。将接枝产物与海藻酸钠溶液混合可以制备均匀的水凝胶。此外，本研究制备的水凝胶显示出理想的抗菌活性。我们发现这些结果十分有趣，我们决定将这篇文章分享在著名杂志《化学通讯》上。

　　如果能得到您的帮助和支持，我们将不胜感激。
致以最良好的祝愿。
诚挚的

张麦克
河北大学副教授
电话：86 139××××
邮箱：××××@hbu.edu.cn
地址：中国河北省保定市五四东路 180 号

Unit 28
Submission

论文投稿（submission）是发表学术论文过程中的一个重要环节。随着互联网通信技术的发展，论文投稿的形式已由原始信件邮寄的方式，转变为经由网页投稿系统进行论文投稿的形式（Fig.2-1，Fig.2-2）。目前，世界范围内的学术交流日益增多，且绝大多数的 SCI、EI 检索期刊的使用语言为英文，因此目前主流的论文投稿系统也均以英文为主。掌握和熟悉英文投稿系统的相关流程、词汇，对于学术论文的撰写、构思和相关材料的准备具有十分重要的意义。

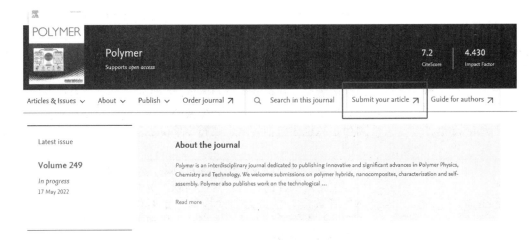

Figure2-1　*Polymer* 期刊网站截图

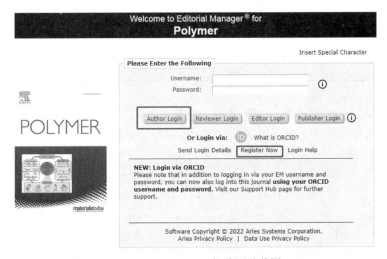

Figure2-2　*Polymer* 投稿网站截图

以 *Polymer* 期刊官网为例，点击"submit your article"选项，即可前往 Editorial Manager® 投稿系统。在该网页中可以通过点击"Register Now"注册一个账号，然后通过输入账号（Username）和密码（Password）点击 Author Login 登录投稿系统。

在投稿过程中需要录入、选择或者上传的信息主要包括：

项目	含义
article type	稿件类型，目前较为常见的几种类型分别是：研究性论文（article/research article/regular article）、综述（review）、快讯（communication/letter）
enter title	输入"题目"
add/edit/remove authors	完善作者信息，一般至少包括作者名字、邮箱、单位
submit abstract	输入"摘要"
enter keywords	输入"关键词"
select classifications	选择分类，一般为稿件涉及的研究方向
enter comments	输入"意见"，此处的意见一般是对编辑部或者期刊的意见，建议谨慎填写
request editor	选择方向相关的主编/编辑
suggested reviewers	推荐审稿人，一般应为稿件相关方向的权威专家，应包括审稿人的姓名、邮箱、单位等信息，某些期刊要求填写推荐理由
opposed reviewers	屏蔽审稿人，因涉及学术保密等问题，建议屏蔽审稿人，所提供信息也应包括姓名、邮箱、单位等信息，以及屏蔽审稿人的理由
information	信息页，该项涉及的内容项目较多，不同期刊要求也不同，主要包括对稿件的一些常规性问题，例如：是否涉及动物伦理？作者数量？是否为初次投稿？是否涉及作者/单位的利益冲突？是否为邀稿？是否选择投稿特刊/专刊等。某些期刊的投稿系统中，cover letter 文件的上传也在此处完成
funding	基金资助，如论文涉及项目资助的声明，应在此处填写信息，一般应包括基金号、资助单位和资助人的信息
attach files	上传文件（重要），此处一般应包括正文（manuscript）、支持信息（supporting information）、图文摘要（graphical abstract）、图片（figure）、图表（table）等文件

在投稿最后阶段，一般有生成 pdf 选项。投稿人一定要下载 pdf，这是由于系统生成 pdf 文件过程中，容易出现字体、格式的改变，使 pdf 版本与原始版本出现偏差。所以我们在下载 pdf 文件后，应在反复确认全部信息无误的情况下，再点击确认投稿的链接。Editorial Manager® 投稿系统中应勾选"I accept"选项后，点击"approve submission"项确认投稿。投稿生成后，可在投稿系统中对稿件的状态（state）进行查询，状态（state）中常见的类型及相关意义汇总如下：

- submitted/ submitted to journal：指稿件已投出，一般而言是稿件投出后的第一个状态。
- received/ received by journal：指编辑部已收到稿件，此时代表稿件投稿成功，并分派稿件

Unit 28　Submission

号码。一般情况下，编辑部的助理编辑或格式编辑会对稿件投稿材料的完整性进行检查，如发现材料不全或信息不完整等情况，编辑部可能会将稿件退回，让投稿人再次完善投稿信息。注意该处的退回只是格式或者材料方面的完善，不涉及稿件学术性问题，应与修订（revision）进行区别。

- awaiting editor assignment/ editor invited / editor assigned：显示该状态时，助理编辑或格式编辑已经完成了对格式和材料的基本审查，并已分配给主编。
- with editor：主编处理稿件中，主编会对稿件的整体情况作出评判，作出是否进一步送审的决定。
- reviewer(s) invited：表示已经邀请审稿人了，等待审稿人同意。稿件的审稿人编辑可能会在 suggested reviewers 中进行选择，也可能不会。
- under review：审稿人审稿中，此过程一般持续时间较长，编辑给予审稿人的时限一般为 2～4 周，如果意见出现严重分歧，还会再次邀请其他审稿人进行仲裁。
- required reviews completed：审稿人审稿意见已返回，等待编辑处理。
- revision / minor revision / major revision：修改/小修/大修。收到修改意见并不代表编辑一定会接收，作者应根据审稿人和编辑的意见对稿件逐条进行修改，以提高稿件质量，并改正不当之处。
- revision submitted to journal：修改稿重新提交，稿件开始新的循环。修改后的论文再次投稿后，编辑有可能会直接作出决断，也可能会将稿件再次送外审，通常会交给第一轮的审稿人，但也有之前审稿人失联，更换审稿人的情况存在。结果如何是根据作者针对评审和编辑意见进行的修改还有回复来决定，如果作者没有完整回复所有的意见，那就要再次修改，甚至有可能会拒稿。
- accepted：稿件接受，待发表。
- reject：拒稿。
- reject and resubmit：拒稿后重新投稿。一般是指稿件存在重大的技术或者方法等问题，这些问题通过大修后，依旧不能很好解决，或者解决这些问题的时间可能比较长，所以编辑做出拒稿的决定。但是编辑或者审稿专家觉得文章还有一定的创新性或者价值，按照专家的意见认真修改后存在可发表的可能，因而鼓励作者按照审稿专家的意见认真修改后，再重新投稿。

接收函示例：

Dear Dr. Zhang,

　　I am pleased to inform you that your manuscript has been accepted for publication in the *Journal Name*.

　　Your accepted manuscript will now be transferred to our production department and work will begin on creation of the proof. If we need any additional information to create the proof, we will let you know. If not, you will be contacted again in the next few days with requests to approve the proof and to complete a number of online forms that are required for publication.

　　Many thanks for submitting your manuscript to the *Journal Name*. We hope that you have been pleased with the handling of your manuscript, and that you will consider submitting other quality manuscripts in the future.

  With kind regards.

Prof. ×××
Co-Editor
Journal Name
Editorial Office

译文
尊敬的张博士：
  我很高兴地通知您，您提交的稿件已被接受，并将在《（期刊名）》上发表。
  您的稿件将被提交到我们的出版部门，并开始制作校样。如果我们需要任何额外的信息来制作校样，我们会联系您。如果没有，我们将在几天后再次联系您，请您确认校样，并填写出版所需的在线表格。
  非常感谢您将稿件投给《（期刊名）》。我们希望您对我们的处理感到满意，并且期待今后您会考虑投来其他高质量的稿件。
  此致
敬礼！

×××教授
联合编辑
《（期刊名）》编辑部

拒稿信示例：
Dear Dr. Li,
  Thank you for submitting your manuscript to be considered for publication in *Journal Name*. After a careful consideration of your manuscript, I regret to inform you that I am unable to accept your manuscript for publication.
  We have carefully examined your manuscript. The demonstrated polymer composites are aimed at treatment of pollutants. Our journal's scope does not include removal of pollutants, unless the materials also have clear applications in other areas that promote biomedical polymer. Hence, we regret to inform you that we are unable to consider the manuscript for publication on *Journal Name*.
  Thank you again for your interest in *Journal Name*.

  Sincerely,

×××
Editor-in-Chief, *Journal Name*
University of Cambridge, UK
Email: ×××
Phone: ×××

Unit 28　Submission

译文

尊敬的李博士：

　　感谢您提交稿件并考虑在《(期刊名)》进行出版。在仔细审阅了您的稿件之后，我很遗憾地通知您，我们不能接受您的稿件出版。

　　我们仔细检查了您的稿件。稿件中提供的聚合物复合材料旨在用于处理污染物。我们杂志的范围不包括污染物的去除方向，除非这些材料在其他促进医用高分子材料的领域也有明确的应用。因此，我们遗憾地通知您，我们不能将您的稿件发表在《(期刊名)》。

　　再次感谢您对本刊的兴趣。

诚挚的
某某某
《(期刊名)》主编
剑桥大学，英国
邮箱：×××
电话：×××

英文稿件投稿注意事项：

- 避免抄袭和学术不端，目前的投稿系统将在投稿初期对稿件进行查重，如果重复比过高，将会直接拒稿。
- 严禁一稿多投。
- 投稿前应对投稿须知（instruction for authors）进行仔细阅读，按照投稿模板或者相关格式要求，准备投稿文件。
- ACS、RSC 的投稿系统账号为数据库通用账号；Wiley、Springer、Elsevier 数据库的投稿系统账号目前仅为期刊内通用，数据库中不同期刊不通用。
- 尽量使用教育网内部邮箱作为投稿邮箱。
- 了解期刊的关注领域（scope），要确定稿件的研究内容在期刊的关注领域范围之内。
- 投稿前需对拟投稿期刊的收费信息进行确认，是否存在出版费或开放获取（open access）费用。
- 了解期刊推荐审稿人的原则。一般而言，具有合作经历的学者不宜推荐；相同工作单位的不宜推荐；同一国家的不宜推荐。
- 了解西方的假期日期，避免假期投稿。
- 谨慎催稿。
- 推荐使用投稿账号关联 ORCID（open researcher and contributor ID）。

Unit 29
Academic Meeting

　　学术会议（meeting）是学术研究的重要日常工作之一。现代社会发展迅速，交通越来越便捷，互联网技术也在不断发展进步，这就使得国际学术会议成为了目前许多高校师生常见的沟通平台。很多大学老师每年都去参加一两个固定主题的年会，一方面增进学术交流，提高自身水平；另一方面也顺便去看望老朋友一样的专业同行（peer researcher），很多学者把某个学术协会或年会称为自己的"academic home"，由此可见学术会议的重要性。meeting 一词泛指会议，规模可大可小，层次可高可低，甚至可以指非正式的聚会。例如 2021 年美国化学会的春季年会，就以"ACS Spring Meeting 2021"命名。

　　除 meeting 一词外，还可利用其它单词将学术会议进一步细致划分。
- conference：一般指大会，使用范围较广。例如 International Conference on Polymer Science。
- congress：应用于学术会议时主要指代表或委员等出席参加的组织范围内的正式会议。
- symposium：研讨会，一般指主题性较强的小范围学术会议，例如 International Symposium on Hydrogel。
- seminar：讨论会，一般用于班级内部、系所内，或课题组内部等的小型专题讨论会。

　　对于一个常规的学术会议而言，一般包括开幕式（general assembly）、全体大会（plenary session）、分组会（parallel sessions）、口头报告（oral presentation）和墙报展讲（poster session）等环节。熟悉不同环节中的常见表达，对于理解会议内容、提高交流效果等方面具有十分重要的意义。

开幕式（general assembly）中的常见表达

- My president of the congress and distinguished guests, ladies and gentlemen, I am very pleased and honored to declare the conference open.
  尊敬的大会主席，尊敬的各位来宾，女士们，先生们，我非常高兴和荣幸地宣布大会开幕。

- The history of the conference dates back to September 1992, when the development of polymer hydrogel has become so important that organizing a meeting is obviously necessary.
  此会议的历史可以追溯到 1992 年 9 月，当时聚合物水凝胶领域的发展已经变得如此重要，组织一次会议显然是必要的。

- The objectives of this meeting are threefold. First, it must provide a forum for participants to exchange information at this interdisciplinary meeting. Second, it must provide an opportunity for participants to revisit old relationships and make new friends. Finally, it must stimulate the interest of participants from all over the world to cooperate as much as possible.
  这次会议有三个目标。首先，它必须为与会者提供一个平台，在这次跨学科会议上交流信

Unit 29　Academic Meeting

息。其次，它必须为参与者提供一个重温旧谊、结交新友的机会。最后，它必将激发世界各地参与者产生尽可能多的合作兴趣。

分组会（parallel sessions）中的常见表达

- I am Zhang Macro. I am from Hebei University, China and I am going to be the chairman for the afternoon's session.
  我是来自河北大学的张麦克，我将担任下午会议的主席。
- There are some basic rules. Each oral report is limited to 15 minutes, and there is 5 minutes for discussion after each new report.
  有一些基本要求需要说明，每个口头报告限时 15 分钟，在每个报告后有 5 分钟的讨论时间。
- There were some changes in the presenters listed in the program. The oral report originally scheduled to be given by professor Zhang was cancelled because he could not attend the meeting. Dr. Li will give a report instead of Professor Zhang.
  列在会议日程上的报告人有些变动，原定由张教授作的口头报告，因为他不能出席这个会议所以取消了。由李博士代替张教授做报告。
- The first report this morning is provided by Professor Zhang from College of Chemistry and Environmental Science, Hebei University. The title of his report is the latest research progress of biomedical polymer materials.
  今天上午第一个报告由河北大学化学和环境科学学院的张教授提供。他的报告题目是医药用高分子材料的最新研究进展。
- Sorry, the schedule is very tight. We haven't any time for discussion, so we must go on to the next report.
  对不起，议程十分紧张，我们没有时间讨论了，所以我们必须继续下一篇报告。

讨论和提问环节（discussion and question）中的常见表达

　　（1）会议主持人关于提问的常见表达

- Time for discussion, do you have any questions?
  讨论时间到了，大家有什么问题吗？
- Any more comments or questions for Dr. Zhang?
  大家还有评论或者其他问题要咨询张博士吗？
- Let's turn to the next problem.
  让我们进行下一个问题。
- I think that will be the last question before we go on to the next oral report.
  这将是下一个口头报告前提问的最后一个问题。

　　（2）提问的常见表达

- Dr. Li, can you further explain the detailed reaction mechanism on this polymerization system?
  李博士，您能否进一步解释一下这个聚合体系的详细反应机理？
- Could you tell us why you use a mixed solvent containing tetrahydrofuran and water to perform the Suzuki coupling reaction?

- 您能告诉我们为什么用四氢呋喃和水的混合溶剂进行铃木偶联反应吗？
- What is your proof that the addition of the synthesized compound can control the rate of polymerization?
  您有什么证据证明添加这种合成的化合物可以控制聚合速率？

（3）回答的常见表达

- Thank you for your question. My answer to that question would be that the use of mixed solvent containing tetrahydrofuran and water in Suzuki coupling reaction is beneficial to avoid the insoluble products with high molecular weight.
  谢谢你的提问。我对这个问题的回答是，在铃木偶联反应中使用含有四氢呋喃和水的混合溶剂有利于避免高分子量的不溶性产物。
- Unfortunately, I can not answer that question on this moment. As far as I know, no enough study has been made on this issue.
  很遗憾，我现在无法回答这个问题。据我所知，在这个问题上还没有充足的研究。
- My colleague, Dr. Yang, may know more about this question. Perhaps she has some better explanation and ideas on this matter.
  我的同事杨博士可能对这个问题了解得更多。也许她对这个问题有更好的解释和想法。

口头报告（oral presentation）中的常见表达

- Thank you, Prof. Lee. Firstly, let me express my appreciation for your kind introduction.
  谢谢 Lee 教授。首先，让我对您十分友好的介绍表示感谢。
- I am very honorable to give this report at ACS Spring Meeting about surface chemical modification of halloysite nanotubes.
  我很荣幸在美国化学会春季年会上就埃洛石纳米管的表面化学改性做报告。
- Today, I am going to summarize some recent advances in the field of biomedical polymers this afternoon.
  今天下午，我将总结生物医学聚合物领域的一些最新进展。
- I would like to present the result of studies on super long persistent materials performed with hydrothermal synthesis method. This work was completed by Prof. Yang, Dr. Wu and me.
  我想介绍一下水热合成法合成的超长持久材料的研究结果。这项工作由杨教授、吴博士和我共同完成。
- I would like to divide my talk this afternoon into three parts. First, synthesis of hyperbranched glycopolymer via a Ce(Ⅳ)-initiated redox system; second, control of molecular weight and degree of branching; and third, the application in inhibiting amyloid agglutination.
  我想把今天下午的演讲分为三个部分。首先，通过铈（Ⅳ）引发的氧化还原体系合成超支化含糖聚合物；第二，控制分子量和支化度；第三，在抑制淀粉样蛋白凝集方面的应用。
- That is the whole content of this report. Thank you for listening. If you are interested in my research work, you can further communicate through email.
  以上就是本次报告的全部内容，感谢各位的倾听。如对我的研究工作感兴趣，可以通过邮箱进一步交流。

Unit 29 Academic Meeting

口头报告（oral presentation）全文讲稿示例

Program Area: PMSE
Session Code: PMSE004A
Symposium Title: Clay/Polymer Composites: Nanoclays and Other Natural Nano-products
Session Title: Nanoplate systems
Session Type: Oral
Session Start Time: Wed 4/07/2021 5:00 PM
Session End Time: Wed 4/07/2021 8:00 PM

Surface Chemical Modification of Halloysite Nanotubes (ID: 3531132)

Thanks to Prof. Zhang, for your kind introduction and help. Good morning everybody, my name is Zhang Hailei, coming from Hebei University. Today my presentation will cover the following aspects: Surface chemical modification of halloysite nanotubes.

As a natural product, halloysite nanotubes exhibit many attractive advantages including good biocompatibility, good solubility in water and environmental friendliness. Our group has paid close attention to halloysite for several years. The TEM and SEM images in this paper show the micro morphology of halloysite purified by our group. It can be suspended into the water uniformly and maintain for more than two days. Benefiting from the tubular structure, halloysite has been widely investigated in carrier material field. It should be noted that the halloysite shows superior character peculiarities when using as drug carrier or zymophore owing to the larger cavity volume and higher biocompatibility. Moreover, the higher thermostability and polyhydroxy character make the HNTs a desirable candidate in heat protection material fields. Furthermore, halloysite also shows promising applications in the field of catalysis, absorbent, energy storage, gel preparation, sensors and so many others. It should be noted that the surface chemical modification of HNTs is the essential precondition to open a broader application, especially for smart materials.

Therefore, in latest years, we devoted ourselves to explore the surface chemical modification methods on halloysite and thereby expand their applications in the field of stimuli-responsive materials, fluorescence probe, antibacterial agent, and so many others. Today, I will briefly introduce three methods about the surface chemical modification of halloysite by using arylboronic acid, supernormal valence transition-metals and silane coupling agent.

Now, let us pay attention to the topic of arylboronic acid. The reaction between arylboronic acid and pinacol is one of the most typical cases in organic chemistry. Moreover, the diboronic acid can be easily developed as an interlinker aiming to developing graphene oxide nanoporous materials with better gas adsorption properties in literature. Therefore, we demonstrate here that the arylboronic acid can also be covalently linked to the alumina innermost surface of halloysite. Following this way, the plane-conjugated molecules pyrenylboronic acid was immobilized onto the lumen surface to functionalize the modified halloysite with fluorescence properties. The product (HNTs-PY) also shows desirable water dispersibility and displays a high fluorescence quantum yield at 0.63. Interestingly, the addition of hydrogen peroxide into HNTs-PY aqueous dispersion triggered a marked decrease in fluorescence intensity. The response mechanism is summarized in Scheme ×××. When pyrenylboronic acid immobilized on to the alumina innermost surface, the nanotubes with pyrene groups can also retain the desirable dispersibility in aqueous media and possess attractive fluorescence property emitting at 380nm. The aromatic boronic acid derivatives can transform into hydroxypyrene in a few minutes in the presence of H_2O_2 owing to the breakage of the B—C link. The poor water solubility of 1-hydroxypyrene results in a dramatically decrease of the fluorescence intensity. The TGA and FTIR results match well with the proposed hydrogen peroxide -responsive character. So, we imagine that the responsive character may be useful in other fields. It should be noted that the halloysite shows attractive advantages when using as drug carrier owing to the larger cavity volume and higher biocompatibility. The halloysite drug delivery system may achieve sustained-release property. Moreover, the drug stability, dispersibility and bioavailability may also be improved. Hydrogen peroxide is one of the most important reactive oxygen species involving a variety of pathological effects in organisms. It has been reported that the aberrant production of H_2O_2 is connected with various diseases, such as cancer, diabetes and cardiovascular. The overexpression of H_2O_2 in inflammation recurrence was observed and has drawn tremendous attention in clinical, especially in the stage of wound healing after operation. Since arylboronic acid modified-HNTs exhibit specific H_2O_2-responsiveness, it spurred us to explore HNTs-based H_2O_2-responsive drug delivery system that may be employed as topical preparations to prevent inflammation occurring. For this purpose, the halloysite was treated with 1,4-phenylenebisboronic acid (PA) giving the product HNTs-BO containing arylboronic acid units. HNTs-BO can be used as a crosslinker based on the reaction of the arylboronic acid and hydroxy group in polysaccharide to give a halloysite-crosslinked hydrogel (HCH). These materials could further be developed as the drug delivery systems via two different approaches. For DLHCH-1, the model drug (pentoxifylline, PTX) was initially loaded into the lumen of HNTs-BO. For DLHCH-2, the hydrogel was achieved by adding the drug into the reaction medium before gel formation, in which the model drug was dispersed in the networks rather than in nanotubes. As shown in Fig.×××, the hydrogel immersing in water retained the self-standing character, and in contrast the hydrogel gradually dispersed into the H_2O_2 solution, which can be owing to the B—C linkage in the obtained hydrogels degraded into boric acid and phenolic hydroxyl group. The results demonstrate the H_2O_2-responsiveness of the hydrogel. The drug loaded hydrogels exhibited the H_2O_2-responsive release behaviour. Moreover, DLHCH-1 showed a much better inhibition effect on "initial burst effect" than that of DLHCH-2, which can be attributed to the loading effect of halloysite. The "initial burst release" is regarded as a puzzling problem in the diffusion-controlled drug delivery systems. So, the halloysite may support an alternative way to solve the puzzling problem. After we have done these works, a problem came to our eyes that the lack of an indicator in the H_2O_2-responsive drug delivery system resulted in a difficulty to monitor the release behaviour in real-time, which may bring some unexpected delays in practical clinical uses. It will be very interesting to further develop a more intelligent H_2O_2-responsive materials that the drug release can be monitored in real-time by the fluorescence changes, which can effectively enhance the efficiency in diagnosis and treatment. For this purpose, a fluorescein derivative bearing two arylboronic acid groups was synthesized and used to react with the hydroxy groups on the halloysite and PVA to give the fluorescein-crosslinked hydrogels. The change from hydroxy group to boronic acid would force the platform to non-fluorescent lactone form, which gave rise to non-fluorescent hydrogels. The addition of H_2O_2 would break the B—C link, trigger the conversion of boronates to phenols, and thereby result in the formation of fluorescein which exhibits a much higher fluorescence quantum yield. Meanwhile, the breakage of B—C linkage would result in the degradation of the hydrogel and give rise to the H_2O_2-responsive drug release behavior. The "turn on" behavior in fluorescence may be quite helpful to trace the responsive release behavior by visual sense. Moreover, a good linear relationship was achieved between the release rate and fluorescence intensity. Just like before, the "initial burst effect" was also be well suppressed. This approach provides a promising opportunity to achieve a new generation of H_2O_2-responsive preparations with visible fluorescence changes to trace the inherent release behavior.

Now, let us pay attention to the second topic. The surface modification of halloysite by supernormal valence transition-metals. It is generally accepted that the supernormal valence transition-metals such as trivalent copper, divalent argentum and tetravalent ceric can efficiently react with alcohol hydroxyl groups in organics and then generate free radicals that may be capable of initiating the vinyl polymerization. In a previous study by our group, the self-condensing vinyl polymerization of carbohydrate-based vinyl monomers was performed by using tetravalent ceric as initiator to afford the hyperbranched glycopolymer with high degree of branching. As a plenty of —OH groups exist on the lumen, there is a question that what will happen by treating HNTs with tetravalent ceric. So, it encouraged us to explore a redox-initiating system that consists of the hydroxyl groups on HNTs and tetravalent ceric, in which the free radicals would be generated on the surface and thereby be capable of initiating a vinyl chain polymerization on HNTs in one step. Fortunately, the redox system of Ce(Ⅳ) and —OH groups located on HNTs was developed to graft the polymer onto HNTs. We named the product as HNTs-EP. The structure is shown in Scheme ×××. The grafting reaction can be performed in aqueous solution at mild temperature. A novel HNTs-based hydrogel was prepared by treating HNTs-EP with sodium alginate in aqueous phase. The figures displayed in ××× show the polymer-grafted HNTs and the obtained hydrogel. The tubular substances can be clearly observed. The product was well characterized by FTIR, NMR and XPS. The results matches well with the proposed mechanism. The grafting degree can be controlled by adjusting the concentration of tetravalent ceric. Moreover, benefiting from the anti-bacterial activity of the polyphosphonium unit, the hydrogel showed an effectively inhibition action upon kanar Gram-negative *Escherichia coli* (*E. coli*). Finally, the cavity of the HNTs in the hydrogel may be used to load drug to achieve a synergistic antibacterial effect accompanying with polyphosphoniums. The redox system established in this work may contribute to a universal method to prepare HNTs-based polymer composites and open up broader applications.

Now, let us go to the third topic. The using of silane coupling agent is a commonly used method to achieve the modification of nanoparticles, of course also including halloysite, which has been reported in several studies. In our work, we focus on developing the facile tools for selective detection and separation of ions. The increasing risk of heavy metals in water course has become a nonnegligible problem. Zinc is making a significant contribution to such pollution, as it has been proved to be rich in wastewater. The excessive amount of zinc in water can eventually lead to the accumulation of zinc in humans, which is directly associated with diabetes and neurodegenerative. So, the removal of zinc from natural waters, as well as the accurate and sensitive detection of zinc ions, has considerable impact. Herein, the aminated halloysite was prepared by treating halloysite with APTES. The aldehyde-containing coumarin derivative was synthesized and anchored on the aminated HNTs via the condensation of aldehydes and amides to afford the Schiff base-containing product (CHNTs). Interestingly, the addition of Zn^{2+} to the CHNTs solution triggered a very quick "turn on" response on fluorescence, accompanying with the appearance of precipitation which was formed by the chelation of Zn^{2+} and CHNTs. The product was well characterized by NMR, and the results match well with the proposed mechanism. In addition, CHNTs displayed excellent selectivity toward Zn^{2+} ions over other metal cations. The elemental mapping results indicated the presence of the zinc in the aggregates of the nanotubes, implying the aggregate may be caused by the zinc. Moreover, the selective removal and detection of Zn^{2+} ions from an aqueous solution containing various metal ions can be facile. The as-prepared CHNTs showing high efficiency and selectivity on the detection and separation of Zn^{2+} will pave the way towards novel approaches for the remediation of heavy metal pollution.

In addition, we also employed some other methods to achieve the surface modification on the halloysite. The reaction between isocynate with hydroxy groups on halloysite was used to functionalize the vinyl group on halloysite to explore it as crosslinking agent. The click chemistry and ATRP route were also used to graft conjugated polymers and ionic liquid onto the halloysite respectively.

The above is the content of today. The photograph shown on ××× is Prof. Ba commissioned me to convey his regards to you. We also thank Prof. Yuri Lvov, who giving us so much help in the recent years. We always welcome everybody here to visit our university. That's all, thank you for your attention.

（以上内容来源于张海磊副教授在 2021 美国化学会春季年会做口头报告的讲稿原文）

Unit 30
Safety

　　安全（safety）是一切工作的前提，对于高分子相关专业和产业而言，安全问题更是重中之重。就学生而言，阅读《安全守则》是进入实验室前的必要环节。《安全守则》中一般会对实验室日常的衣着穿戴、试剂购置、药品存放、废液/固处理、易燃易爆注意事项、剧毒品、实验卫生等诸多注意事项进行一一说明。国外实验室同样十分重视安全问题，这就要求同学们在进入实验室前，一定要熟读英文版的《安全守则》，即 Safety Guideline。

　　此外，关于高分子实验室和企业生产环境中有关安全的标志，其中英文相关含义，也是需要同学们掌握和熟记的重点。这不仅是对实验工作的重视，更是为自己的生命健康负责。

常用的符号：

	flammable	易燃的
	harmful	有害的
	serious health hazard	严重危害健康
	explosive	易爆/易爆品
	oxidising	具有氧化性的
	toxic	有毒的
	environmental hazard	环境有害的
	corrosive	腐蚀性的/腐蚀物
	gas under pressure	高压气体
	radiation risk	辐射危险

（续表）

	harmful or irritating substances	有害或刺激性物质
	UV light	紫外光
	dangerous electric tension	有电危险
	explosive atmosphere	易爆气体环境
	poisonous substances	毒性物质
	high intensity magnetic field	强磁场
	explosive substances	易爆物
	corrosive substances	腐蚀物
	inflammable substances	易燃物
	biological risk	生物危险性
	hanging loads	悬挂荷载
	oxidizing substances	氧化物
	laser ray	激光射线
	low temperature	低温
	danger	危险
	hot surface	热表面
	suffocation	窒息
	obliged to wear safety glasses	必须戴安全眼镜

（续表）

	obliged to wear safety gloves	必须戴安全手套
	obliged to wear overshoes	必须穿鞋套
	obliged to wear hearing protection	必须佩戴听力保护装置
	obliged to wear safety shoes	必须穿安全鞋
	obliged to wear countenance protection	必须佩戴防护面罩
	obliged to wear breath protection	必须佩戴防毒面具
	obliged to wear safety helmet	必须戴安全帽
	obliged to wear breathing protection	必须佩戴口罩
	first aid	急救
	eye shower	洗眼器
	emergency shower	紧急喷淋装置
	warning button	报警按钮
	fire extinguisher	灭火器
	fire hose	消防水带

示例 1

Basic Lab Safety and General Rules

Every co-worker is considered to be fully aware of the following specific info and safety rules within the laboratory:

GENERAL ISSUES:

- In case of emergency. On the first floor of the building, there is a first-aid room for people not feeling well or for injuries. There are first-aid kits in the main laboratory also on the 1st floor. Make sure that you know where the first-aid kits are stored and where the first-aid room is.
- If the evacuation alarm goes off, you have to leave the building immediately. Do not use the elevators, use stairs only. Go to the assembly point and wait there at the correct floor assembly point.
- Make sure that you know where to find the nearest fire extinguishers, exits, emergency showers, eye showers, fire blankets, first aid kits and telephones.
- You have to report all incidents and accidents as soon as possible to Dr. Zhang for further follow-up.
- All equipment problems should be reported to Dr. Zhang or the lab responsible.
- If you are not sure how to perform an experiment or use an instrument, always ask for help. There is a list of equipment/responsible next to Dr. Zhang office.
- Please note that the lab computers are for work use only. Do not use them for private business; the same applies for printers and copiers. Illegal downloading can & will be tracked back to the specific computer & user who will personally be held responsible.
- For most of the instruments (UV/Vis spectrophotometer, fluorescence spectrophotometer, GPC, etc.), you have to "book" the instrument for the time you want to use it (write your name and date on the reservation paper next to the instrument).
- Make sure that you re-order all lab products in the stock cupboard when the re-order amount (second to last bottle) has been reached or inform to Dr. Zhang.
- If you finish a chemical, order a new bottle. Use the program to order all chemicals and print the order forms, except for solvents.
- General lab responsible is Dr. Zhang. He will supervise and ensure that everybody follows the rules regarding health and safety in the lab.
- Every last Friday of every month, a general lab cleaning will be held. Every week there will be one person responsible for taking out the solid and liquid waste, filling the water and acetone tanks and making sure everything is switched off in the evening. Please check the lab responsible list on your floor and follow the instructions of that document.

SAFETY ISSUES:

- You have to wear safety glasses, lab coat, long trousers (e.g. legs must be fully covered) and fully-covered shoes at all times. Nylon stockings are forbidden for safety reasons and long hair should be tied.
- For overnight reactions and large scale reactions (> 50g reagent or > 200mL), the Safety Information Sheet should be filled out and put on your fume hood. The safety sheet must be signed by a Postdoc only.
- For rotary evaporation of corrosive organic/inorganic compounds, please used the dedicated rotary evaporator located on the 2nd floor lab. Also, large scale reactions must be performed on that lab.

Unit 30 Safety

- Drinking and eating are not allowed in the labs. Don't store food and drinks in the labs.
- You are not allowed to bring chemicals into the computer office, relax room. Before you leave the lab, you have to remove gloves and lab coat and wash your hands. Especially the relax room has to be kept completely chemical-free as food is prepared here. Also remove gloves before you answer the phone.
- You are not allowed to work in the lab alone; at least one fully trained person (= PhD, postdoc, technician) must be on the same floor beside yourself.
- You have to wear gloves when working with chemicals. Please note that latex and nitrile gloves are not leak proof for most solvents, e.g. they will not protect you if solvents get on the gloves.
- Take off your lab coat and gloves when leaving the lab to avoid contaminating doors and handles.
- You have to label all your flasks and samples with your initials, the name of the chemical and if necessary danger symbols/text (e.g. flammable, explosive,…). DO NOT re-label a bottle containing a different chemical.
- For longer stocking, do not use general glassware such as flasks. There are plastic bottles of appropriate size available for this.
- When working with azides, fill in the laboratory safety sheet. Sign the form after reading and give it to Dr. Zhang. Same applies to freeze-pump-thaw cycles and its protocol.
- When working with freeze-pump-thaw cycles, carefully read and follow the special protocol. Take special care not to condense gases during this procedure!
- You are not allowed to work with hazardous chemicals/solvents on the bench. Use a fume hood while working with hazardous chemicals.
- The drying lines may only be used to remove traces of solvent in already dry samples.
- Small particles are dangerous for the respiratory system and should therefore be handled in the fume hood.
- Do not store filled NMR tubes on your desk. Keep them in your fume hood as the NMR-solvents will slowly evaporate. NMR tubes are very expensive and are not suitable for storing your products. They should be emptied and cleaned immediately after use. Rinsing the NMR tubes with chloroform will help them dry better. A long needle can be used to dry them more sufficient. Do not reuse damaged NMR tubes.
- Keep the windows of the fume hood as low as possible at all times. Work behind the blast shield as much as possible when doing an experiment which involves high vacuum e.g. freeze-pump-thaw cycles.
- Chemical waste has to be segregated, e.g. flammable non-halogenated, flammable halogenated, solid waste, glass waste, chemical contaminated waste, and others. Please pay attention to collecting things in the correct waste barrels. High fees exist for not-appropriately collected waste. Always leave tissue, syringes with chemicals "dry" overnight in the fume hood before throwing in the chemical contaminated WASTE bins. If you have a doubt, ask first.
- Do not throw glass in the chemical contaminated WASTE bins.
- Empty chemical bottles (< 1 liter) have to be put in the chemical contaminated GLASS bin. Do not put 'normal' glass waste (drinking bottles) in the chemical contaminated glass bin. Empty

chemical bottles (> 1 liter) have to be placed under the oven. They have to be brought to the waste room in the basement every Friday by the corresponding lab responsible of the week. If the empty bottle is being reused for a different purpose, please remove the barcode and place it on a small sheet of paper under the oven. This way the chemical can be removed from the database and "leave the building".

- Needles have to be put in special needle containers. Store needle containers in fume hood. When they are full, bring them to the waste room in the basement.
- Do not throw any chemical contaminated waste in the normal bin.
- Do not clean glassware that is contaminated with hazardous chemicals in the sink; clean them in the fume hood to avoid bad smells in the lab!
- Chemicals are not to be stored on lab benches or in lab cupboards but in the stock room. After using a chemical, return it to the stock room as soon as possible. Keep the door of the stock room closed and locked.
- All waste containers have to be capped immediately when full, and bring them to the chemical waste room in the basement. Make sure to order new waste containers in time via the order form.
- Do not use methanol to clean glassware (methanol is very toxic!), use acetone.
- You have to clean up immediately after you have finished your experiments.
- Close the door of the analytical balance after use. Open door can harm the balance! Make sure that you clean the balance after use.
- Please close the door when you are the last person leaving your lab.

示例 2

Use of Hazardous Substances

Hazardous substances are stored in the original, non-expired packaging, provided with the correct labels and danger symbols. Anyhow, when moving substances from one container to another, the receiving container has to meet the required rules of packaging and carry the labels and symbols required by regulation.

The Safety Data Sheets or SDS provide this information more extensively. The supplier is legally obliged to deliver an updated SDS for each supplied substance. In practice, unfortunately, the user often needs to claim it himself. The Safety Data Sheets of most common used substances are available in the hazardous substances database on Apollo. For certain specific substances the users have to provide these sheets themselves. The Safety Data Sheets must be accessible to the staff of the department at any time. Hazard labels can be printed from the Hazardous Substances Database.

The use of hazardous substances requires an evaluation of the potential risks to avoid or minimize exposure to humans and to the environment. Potential risk should be summarized in an A4 paper and posted onto the doors of each laboratory as well as the telephone numbers of contact persons. The emergency instructions, the manual of the lab and specific standard operation procedures should always be available in the lab, as well as the SDS of the handled products or the technical sheets of the waste streams produced.

The presence of hazardous substances in laboratories and workplaces should be limited. In practice, this means only those amounts necessary for ongoing experiments should be present. At the end of the day, inflammable substances must be stored in an appropriate storage depot, so called group 1-room or a safety cupboard. Very toxic substances should be stored immediately after the activities in a locked area. It's forbidden to store more than 50 L of inflammable liquids in a room, if this room is not equipped in accordance with storage room requirements. Storage of hazardous substances in a lab is not allowed. They should be stored in a special room equipped in accordance with storage room requirements. If such a storage room is not available, or if the quantities are very limited, the storage of inflammable substances may be allowed in safety cupboards. Other dangerous substances can be stored in a suitable cupboard or shelf. Corrosive substances are best stored in an acid and alkali storage cabinet. Cupboards for liquids must be provided with ledgers with standing boards.

Appropriate storage depots for the storage of hazardous products are storage rooms, cupboards and sheds designed with that purpose. They should comply with the safety and environment regulations. Inflammable substances must be stored in a group-1 room. The rooms must be cleaned up and ventilated. Moreover, products must be classified by danger category with the labels in the front and last but not least, hazardous substances should be put in rooms, cupboards or sheds with standing boards. Don't put these substances any higher than shoulder height. Substances which can react dangerously with others shouldn't be placed together.

Waste production and collection also cause risks, especially in the case of hazardous waste. Waste is categorised in nonhazardous waste and hazardous waste that contains medical, animal, radioactive and small dangerous waste. Hazardous waste is under the supervision of the environment office, except for radioactive waste that is managed by the radiation protection service. Each department of the faculty is responsible for the internal organisation of the selective collection. Waste is only collected in consultation with the services mentioned above, which keep the required waste inventories.

Students and staff know which waste belongs to which waste stream by a clearly marked 'waste corner', provided with waste posters. Technical sheets—available for all present waste streams in a room—give more detailed information about waste containers, labeling, collecting conditions, transport, etc. Filled barrels are brought to storage depots prior to their transport. At the end of the day, inflammable waste is placed in a fireproof storage depot or in a safety cupboard. Anyway, hazardous waste should be transported before the expiry date of the plastic containers (5 years).

In the environmental permission, emission standards for waste water are defined. Exceeding these emission standards is an environmental crime and can lead to a financial penalty. This shall be paid by the producers of pollution, according to the polluter-pays principle. The extra costs for remediation, e.g. extra samplings, will also have to be paid for when the causes are known, but not dealt with.

The most critical parameters:
- heavy metals as Pb, Zn, Cd, Hg, Ag
- organic halogenated solvents as dichloromethane, chloroform

- some poly-aromatic hydrocarbons
- tributyltin
- nonylphenol
- chloramines
- polybrominated diphenyl ethers
- pentachlorobenzene

It's strictly forbidden to discharge these.

Besides the above-mentioned guidelines for hazardous substances, some additional rules are given for a few specific hazardous substances. Because ethidiumbromide is a marker for DNA, it is used by many biotechnological laboratories. Moreover, ethidiumbromide is a very toxic substance (mutagen and carcinogen). For that reason, manipulations and the use of protective equipment are strictly regulated. Use for this work a separate room or—when not available—look for a conform ethidiumbromide room, that can be shared with other research groups. Within the room, the contamination zone should be clearly distinguished from the zone without contamination. For the use of SybrSafe®(American Database Products Inc) and other DNA markers, the same guidelines as for the use of ethidiumbromide apply. SybrSafe has the advantage that no UV-light is required.

Pathogens, hazardous biological agents for human, plant, animal and environment (bacteria, viruses, parasites, fungi, etc.) and genetically modified micro-organisms should be used and stored in accordance with the biosafety regulation. A risk level is pointed according to the biological agents and the activity. Therefore, containment levels count for infrastructure as well as for working practices. A specific permission by the government is required for every biotechnological activity.

Before using radioactive substances or ionizing radiation, contact the Radiation Protection Service and the Department of Occupational Health. A licence of the Federal Agency for Nuclear Control (FANC), a specific waste-procedure, a specific medical examination and often personal dosimeters are required for these activities. Concerning the specific guidelines in each department, please contact the certified person. In labs working with radioactive substances, all desks must be easy to clean and the room must have standing boards.

Liquid nitrogen is one of the most used cryogen liquids. Liquid nitrogen is a gas that has been condensed to a liquid. It has an extremely low temperature of about $-196\,^\circ\!\text{C}$. When this liquid is exposed to outside air, it shifts boiling into its gas form. The transition from liquid to gas triggers a forceful increase in pressure and/or an increase in volume. In its transition to gas, liquid nitrogen's volume increases 700 times. This increase in pressure can lead to explosions. The transition to the gas state also implies that the nitrogen concentrations in the air increase, and that the oxygen concentration, normally amounting to 21%, strongly decreases, heightening the danger of suffocation for people that enter the space at that moment.

模拟测试

模拟测试题（一）

考核科目：高分子专业英语　考试时间：90 分钟　考核方式：闭卷

一、将下列英文单词译成中文（每小题 1 分，总分 20 分）

（一）modification　　　（二）architecture　　　（三）performance
（四）purification　　　（五）polymerization　　　（六）polyethylene
（七）propylene　　　　（八）acetone　　　　　（九）precipitation
（十）aggregation　　　（十一）hydroxy　　　　（十二）parameter
（十三）cellulose　　　（十四）membrane　　　（十五）configuration
（十六）polypropylene　（十七）ethylene　　　　（十八）ethanol
（十九）alkali　　　　　（二十）rubber

二、将下列中文词语译成英文（每小题 2 分，总分 10 分）

（一）*n.* 聚合物　　　（二）*n.* 溶解　　　（三）*n.* 溶剂
（四）*n.* 聚酯　　　　（五）离子聚合（词组）

三、句子翻译（将英文句子翻译为中文，每小题 5 分，共 30 分）

（一）It is essentially the 'giantness' of the size of the polymer molecule that makes its behavior different from that of a commonly known chemical compound such as benzene.

（二）Such reactions occur through the initial addition of a monomer molecule to an initiator radical or an initiator ion, by which the active state is transferred from the initiator to the added monomer.

（三）Reaction (3.1) illustrates the former, while (3.2) is of the latter type.

（四）With ionic polymerization there is no compulsory chain termination through recombination, because the growing chains can not react with each other.

（五）It relies on the introduction of a reagent that undergoes reversible termination with propagating radicals thereby converting them to a following dormant form.

（六）Thus, strength of polymer does not begin to develop until a minimum molecular weight of about 5000-10000 is achieved.

四、段落翻译（将英文翻译为中文，共 20 分）

Hyperbranched polymers have attracted considerable interest because of their unique molecular structures and properties, such as good solubility, low viscosity, and large numbers of functional end groups. A degree of branching (DB) of 100% is a typical feature for perfectly branched dendrimers, whereas hyperbranched polymers prepared by one-pot reaction are usually not perfectly branched and have a DB of ~50%. Many attempts have been made to improve the degree of branching of hyperbranched polymers. These modification methods only enhance the DB, but none of them can achieve a hyperbranched polymer with a DB of 100%. Hyperbranched polymers with a DB of 100% have been recently reported by several groups but are limited to few monomers.

五、写作（20分）

根据高分子化学实验中的聚甲基丙烯酸甲酯（polymethyl methacrylate，PMMA）本体聚合实验，撰写论文的摘要（abstract）部分，字数在150～200字。

写作内容应包含：

实验目的

实验方法概述

表征方法简介（如聚甲基丙烯酸甲酯分子量的表征方法等）

意义及展望

模拟测试题（二）

考核科目：高分子专业英语　考试时间：90 分钟　考核方式：闭卷

一、将下列英文单词译成中文（每小题 1 分，总分 20 分）

（一）molecule　　　　　　（二）swell　　　　　　（三）backbone
（四）ionic　　　　　　　（五）linear polymer　　（六）poly(1-butene)
（七）styrene　　　　　　（八）methanol　　　　　（九）cation
（十）carbohydrate polymer　　（十一）conformation
（十二）reactant　　　　　（十三）amorphous　　　（十四）dendrimer
（十五）reflux　　　　　　（十六）polyisobutylene　（十七）starch
（十八）dichloromethane　　（十九）centrifuge
（二十）mean square end-to-end distance

二、将下列中文词语译成英文（每小题 2 分，总分 10 分）

（一）n. 纤维素　　（二）n. 溶解度　　（三）n. 共聚物
（四）v. 合成　　　（五）活性自由基聚合（词组）

三、句子翻译（将英文句子翻译为中文，每小题 5 分，共 30 分）

（一）Another peculiarity is that, in water, polyvinyl alcohol never retains its original powdery nature as the excess sodium chloride does in a saturated salt solution.

（二）These are energy-rich compounds which can add suitable unsaturated compounds (monomers) and maintain the activated radical, or ionic, state so that further monomer molecules can be added in the same manner.

（三）The net effect of esterification is that monomer molecules are consumed rapidly without any large increase in molecular weight.

（四）If the initiators are only partly dissociated, the initiation reaction is an equilibrium reaction, where reaction in one direction gives rise to chain initiation and in the other direction to chain termination.

（五）This enables the active species concentration to be controlled and thus allows such a condition to be chosen that all chains are able to grow at a similar rate.

（六）When one speaks of the molecular wight of a polymer, one means something quite different from that which applies to small-sized compounds.

四、段落翻译（将英文翻译为中文，共 20 分）

Noncovalent interactions between complex carbohydrates and proteins drive many fundamental processes within biological systems, including human immunity. In this report we aimed to investigate the potential of mannose-containing glycopolymers to interact with human DC-SIGN and the ability of these glycopolymers to inhibit the interactions between DC-SIGN and the HIV envelope glycoprotein gp120. We used a library of glycopolymers that are prepared via combination of copper-mediated living radical polymerization and azide-alkyne [3+2] Huisgen cycloaddition reaction. We demonstrate that a relatively simple glycopolymer can effectively prevent the interactions between a human dendritic cell associated lectin (DC-SIGN) and the viral envelope glycoprotein gp120. This approach may give rise to novel insights into the mechanisms of HIV infection and provide potential new therapeutics.

五、写作（20分）

以自由基聚合（radical polymerization）为主题，撰写引言（introduction）第一部分，字数在150~200字。

写作内容应包含以下信息：
聚合反应的种类
自由基聚合的特点
自由基聚合的优势
自由基聚合成功应用的事例简介

模拟测试题（三）

考核科目：高分子专业英语　考试时间：90 分钟　考核方式：闭卷

一、将下列英文单词译成中文（每小题 1 分，总分 20 分）

（一）nylon　　　　　　　（二）diamine　　　　　　（三）characterize
（四）proton　　　　　　　（五）branched polymer　　（六）polyvinylchloride
（七）1-butene　　　　　　（八）chloroform　　　　　（九）concentration
（十）density　　　　　　　（十一）compression　　　（十二）biodegradable
（十三）mechanism　　　　（十四）polyfunctional　　（十五）data
（十六）polystyrene　　　　（十七）plastic　　　　　（十八）tetrahydrofuran
（十九）protection　　　　（二十）shear modulus

二、将下列中文词语译成英文（每小题 2 分，总分 10 分）

（一）n. 大分子　　　（二）n. 引发剂　　　（三）n. （链）增长
（四）n. 产率　　　　（五）聚合速率（词组）

三、句子翻译（将英文句子翻译为中文，每小题 5 分，共 30 分）

（一）For example, two separate linear molecules have a total of four ends. If the end of one of these linear molecules attaches itself to the middle of the other to form a "T", the resulting molecule has three ends.

（二）Radical chain polymerization is a chain reaction consisting of sequence of three steps—initiation, propagation and termination.

（三）Polyesterification, whether between diol and dibasic acid or intermolecularly between hydroxy acid molecules, is an example of a step-growth polymerization process.

（四）As has been described in detail with radical polymerization, one can characterize each monomer pair by so-called reactivity ratios r_1 and r_2.

（五）Ideally, the mechanism of living polymerization involves only initiation and propagation steps. All chains are initiated at the commencement of polymerization.

（六）The interesting and useful mechanical properties which are uniquely associate with polymeric materials are a consequence of their high molecular weight.

四、段落翻译（将英文翻译为中文，共 20 分）

Conjugated polymers have been extensively studied recently due to their potential applications as light-emitting materials. For achieving full color displays, the three primary-color-emitting materials, i.e., blue, green, and red, are essential. Compared with the high brightness and efficiency of green-and red-light emitting materials, blue-light-emitting material does not match requirements for commercially feasible light-emitting diodes. However, all these materials suffer from the low-energy emission generated during either annealing or passage of current in solid states.

五、写作（20 分）

以逐步聚合（step-growth polymerization）为主题，例如铃木偶联反应（Suzuki-coupling reaction），点击化学（click chemistry）等，撰写引言（introduction）第一部分，字数在 150～200 字。

写作内容应包含以下信息：
聚合反应的种类
逐步聚合的特点
逐步聚合的优势
逐步聚合成功应用的事例简介

模拟测试题（四）

考核科目：高分子专业英语 考试时间：90 分钟 考核方式：闭卷

一、将下列英文单词译成中文（每小题 1 分，总分 20 分）

（一）polydispersity　　　（二）solubility　　　（三）backbone
（四）diacid　　　（五）number-average molecular weight
（六）polyvinyl alcohol　　　（七）acrylamide　　　（八）dimethyl sulfoxide
（九）deprotection　　　（十）element　　　（十一）abbreviate
（十二）segmer　　　（十三）synthesize　　　（十四）radical
（十五）termination　　　（十六）initiator　　　（十七）random coil
（十八）dimethyl formamide　　　（十九）extract　　　（二十）Young modulus

二、将下列中文词语译成英文（每小题 2 分，总分 10 分）

（一）n. 单体　　　（二）n. 均聚物　　　（三）n.（链）引发
（四）n. 转化率　　　（五）分子量（词组）

三、句子翻译（将英文句子翻译为中文，每小题 5 分，共 30 分）

（一）When we speak of a branched polymer, we refer to the presence of additional polymeric chains issuing from the backbone of a linear molecule.

（二）The value of k_p for the most monomers is in the range of 10^2-10^4 L/(mol·s). This is a large constant—much larger than those usually encountered in chemical reactions.

（三）These compounds are prepared from polymerization of tetrahydrofuran; they are also linear, while those prepared from butylene oxide are marked by pendant groups.

（四）Ionic polymerization, similar to radial polymerization, also has the mechanism of a chain reaction. The kinetics of ionic polymerization are, however, considerably different from that of radical polymerization.

（五）Up to now, several living radical polymerization processes, including atom transfer radical polymerization and reversible addition-fragmentation chain transfer polymerization, have been reported one after another.

（六）When one discusses the molecular weight of a polymer, one is actually involved with its average molecular weight.

四、段落翻译（将英文翻译为中文，共 20 分）

The polymerization rate may be experimentally followed by measuring the changes in any of several properties of the system such as density, refractive index, viscosity, or light absorption. Density measurements are among the most accurate and sensitive of the techniques. The density increases by 20-25 percent on polymerization for many monomers. In actual practice the volume of the polymerizing system is measured by carrying out the reaction in a dilatometer. This is specially constructed vessel with a capillary tube which allows a highly accurate measurement of small volume changes. It is not uncommon to be able to detect a few hundredths of a percent polymerization by the dilatometer technique.

五、写作（20 分）

以自己做过的某高分子合成领域相关实验为主题，撰写论文的摘要（abstract）部分，字数在 150~200 字。

写作内容应包含：
实验目的
实验方法概述
表征方法简介
意义及展望

模拟测试题（一）答案及评分标准

一、将下列英文单词译成中文（每小题 1 分，总分 20 分）

（一）改性　　（二）结构　　（三）性能
（四）纯化　　（五）聚合　　（六）聚乙烯
（七）丙烯　　（八）丙酮　　（九）沉淀
（十）聚集　　（十一）羟基　　（十二）参数
（十三）纤维素　（十四）膜　　（十五）构型
（十六）聚丙烯　（十七）乙烯　　（十八）乙醇
（十九）碱　　（二十）橡胶

二、将下列中文词语译成英文（每小题 2 分，总分 10 分）

（一）polymer　　（二）dissolution　　（三）solvent
（四）polyester　　（五）ionic polymerization

三、句子翻译（将英文句子翻译为中文，每小题 5 分，共 30 分）

（一）聚合物分子巨大的尺寸使它的某些行为与常见的化合物存在区别，比如苯。

（二）这类反应是通过单体分子首先加成到引发剂自由基或引发剂离子上而进行的，靠这些反应活性中心由引发剂转移到被加成的单体上。

（三）反应（3.1）说明前一种形式，而反应（3.2）说明后一种形式。

（四）对于离子聚合而言，不存在通过再结合反应而进行的强迫链终止，因为生长链之间不能发生链终止反应。

（五）它依赖于向体系中引入一种可以和增长自由基进行可逆终止反应的试剂，形成休眠种。

（六）因此，直到最小分子量增大到 5000~10000 以后，聚合物的强度才开始显示出来。

（要求内容正确，句子流畅通顺。词语错误或单词翻译错误每个扣 1 分，句子不通顺扣 1 分）

四、段落翻译（将英文翻译为中文，共 20 分）

超支化聚合物由于其独特的分子结构和性质，如良好的溶解性、低黏度和大量的官能团等，引起了人们的广泛关注。100%的支化度（DB）是完全支化的树枝状大分子的典型特征，而一锅法合成的超支化聚合物通常不是完美树枝状分子，其 DB 为~50%。为了提高超支化聚合物的支化度，人们做了许多尝试。这些改性方法只能提高聚合物的 DB 值，但没有一种方法能得到 DB 值为 100%的超支化聚合物。最近一些研究团体报道制取了 DB 值为 100%的超支化聚合物，但仅限于少数单体。（要求内容正确，句子流畅通顺。词语错误或单词翻译错误每个扣 1 分，句子不通顺扣 1 分）

五、写作（20 分）

一档（16～20 分）突出摘要的全部内容要点，包括实验目的、实验方法、表征方法简介和意义等内容。层次清楚，语言流畅，有句式的变化，有复杂结构，基本无语法错误或拼写错误。

二档（11～15 分）基本写出摘要的内容要点，层次清楚，语言流畅，存在少量的语法和拼写错误。

三档（6～10 分）包括了摘要一半以上的内容要点，语言不通顺，语法结构单一，只有少量句子完整无误，影响整体阅读。

四档（0～5 分）不扣题，词不达意，整休无法阅读。

其他要求及评分说明：字数不符合标准扣 2 分，书写潦草影响阅读扣 2 分，语法和拼写错误扣 0.5～5 分。

模拟测试题（二）答案及评分标准

一、将下列英文单词译成中文（每小题 1 分，总分 20 分）

（一）分子　　　　　（二）溶胀　　　　　（三）骨架
（四）离子的　　　　（五）线型聚合物　　（六）聚丁烯
（七）苯乙烯　　　　（八）甲醇　　　　　（九）阳离子
（十）碳水化合物的聚合物/聚糖　　　　　（十一）构象
（十二）反应物　　　（十三）无定形　　　（十四）树枝状大分子
（十五）回流　　　　（十六）聚异丁烯　　（十七）淀粉
（十八）二氯甲烷　　（十九）离心机　　　（二十）均方末端距

二、将下列中文词语译成英文（每小题 2 分，总分 10 分）

（一）cellulose　　　（二）solubility　　　（三）copolymer
（四）synthesize　　　（五）living radical polymerization

三、句子翻译（将英文句子翻译为中文，每小题 5 分，共 30 分）

（一）另一个特点是，在水中聚乙烯醇不会像过量的氯化钠在饱和盐溶液中那样能保持其初始的粉末状态。

（二）这些（化合物）是高能态化合物，他们可以加成不饱和化合物（单体），并且（在完成一步加成以后仍然）保持活性自由基或离子状态便于类似地进一步加成单体分子。

（三）酯化网状效应就是单体分子很快消耗，而（聚合物的）分子量却没有增大多少。

（四）如果引发剂仅为部分地离解，引发反应即为一个平衡反应，该反应在一个方向上导致链引发，而在另一个方向上则引起链终止。

（五）这使活性中心的浓度能够得以控制，由此可以选择适宜的反应条件，使所有的分子链都能够以相同的速度增长。

（六）人们谈论的聚合物的分子量，与小分子化合物的分子量完全不同。

（要求内容正确,句子流畅通顺。词语错误或单词翻译错误每个扣 1 分,句子不通顺扣 1 分）

四、段落翻译（将英文翻译为中文，共 20 分）

糖和蛋白之间的非共价键相互作用驱动了生物系统中很多重要的过程，包括人体免疫。由本文我们旨在研究甘露糖基聚糖与人体 DC-SIGN 的相互作用，和这些聚糖抑制 DC-SIGN 和艾滋病壳膜糖蛋白 gp120 作用的能力。我们使用了一系列通过铜介导的活性自由基聚合和叠氮-炔[3+2]惠斯更环加成反应制备的聚糖。我们证明了这样一个相对简单的聚糖可以有效抑制人树突细胞相关凝集素 DC-SIGN 和艾滋病壳膜糖蛋白 gp120 的相互作用。这一研究给解读艾滋病的感染机制提供了新的思路，也提供了可能的新治疗方案。

（要求内容正确,句子流畅通顺。词语错误或单词翻译错误每个扣 1 分,句子不通顺扣 1 分）

五、写作（20分）

一档（16～20分）突出自由基聚合（radical polymerization）为主题的引言（introduction）第一部分的全部内容要点，包括聚合反应的种类、自由基聚合的特点、自由基聚合的优势、自由基聚合成功应用的事例简介。层次清楚，语言流畅，有句式的变化，有复杂结构，基本无语法错误或拼写错误。

二档（11～15分）基本写出引言（introduction）第一部分的内容要点，层次清楚，语言流畅，存在少量的语法和拼写错误。

三档（6～10分）包括了引言（introduction）第一部分一半以上的内容要点，语言不通顺，语法结构单一，只有少量句子完整无误，影响整体阅读。

四档（0～5分）不扣题，词不达意，整体无法阅读。

其他要求及评分说明：字数不符合标准扣2分，书写潦草影响阅读扣2分，语法和拼写错误扣0.5～5分。

模拟测试题（三）答案及评分标准

一、将下列英文单词译成中文（每小题 1 分，总分 20 分）

（一）尼龙　　　　　　（二）二胺　　　　　　（三）表征
（四）质子　　　　　　（五）支化聚合物　　　（六）聚氯乙烯
（七）1-丁烯　　　　　（八）三氯甲烷/氯仿　　（九）浓度
（十）密度　　　　　　（十一）模压　　　　　（十二）可生物降解的
（十三）机理/机制　　　（十四）多功能的　　　（十五）数据
（十六）聚苯乙烯　　　（十七）塑料　　　　　（十八）四氢呋喃
（十九）保护　　　　　（二十）剪切模量

二、将下列中文词语译成英文（每小题 2 分，总分 10 分）

（一）macromolecule　　（二）initiator　　（三）propagation
（四）yield　　　　　　（五）rate of polymerization

三、句子翻译（将英文句子翻译为中文，每小题 5 分，共 30 分）

（一）例如，两个分开的线性分子一共有四个端基。如果其中一个线性分子的端基连接到另一个的中间，形成一个"T"，得到的分子具有 3 个端基。

（二）自由基链式聚合是一种链式反应，它包括三个步骤：引发，增长和终止。

（三）无论是二元酸和二元醇之间或是羟基酸分子之间的聚酯反应，都是逐步聚合过程的一个典型例子。

（四）正如自由基聚合的详细描述，我们可以通过所谓的竞聚率 r_1 和 r_2 来表征每对单体。

（五）理想情况下，活性聚合的机理只涉及引发和增长步骤。所有的链都是在聚合开始时被引发的。

（六）高分子材料的高分子量决定了其独特的有趣且有用的力学性能。

（要求内容正确，句子流畅通顺。词语错误或单词翻译错误每个扣 1 分，句子不通顺扣 1 分）

四、段落翻译（将英文翻译为中文，共 20 分）

共轭聚合物作为发光材料具有潜在的应用前景，近年来得到了广泛的研究。为了实现全彩显示，三种原色的发光材料，即蓝色、绿色和红色，是必不可少的。与绿色和红色发光材料的高亮度和高效率相比，蓝色发光材料不符合商业上可行的发光二极管的要求。然而，所有这些材料在退火或电流在固态中通过过程中都会产生低能发射。

（要求内容正确，句子流畅通顺。词语错误或单词翻译错误每个扣 1 分，句子不通顺扣 1 分）

五、写作（20 分）

一档（16～20 分）突出逐步聚合为主题的引言（introduction）第一部分的全部内容要点，包括聚合反应的种类、逐步聚合的特点、逐步聚合的优势、逐步聚合成功应用的事例简介。

层次清楚，语言流畅，有句式的变化，有复杂结构，基本无语法错误或拼写错误。

二档（11~15分）基本写出引言（introduction）第一部分的内容要点，层次清楚，语言流畅，存在少量的语法和拼写错误。

三档（6~10分）包括了引言（introduction）第一部分一半以上的内容要点，语言不通顺，语法结构单一，只有少量句子完整无误，影响整体阅读。

四档（0~5分）不扣题，词不达意，整体无法阅读。

其他要求及评分说明：字数不符合标准扣2分，书写潦草影响阅读扣2分，语法和拼写错误扣0.5~5分。

模拟测试题（四）答案及评分标准

一、将下列英文单词译成中文（每小题 1 分，总分 20 分）

（一）多分散性　（二）溶解度　　（三）骨架
（四）二酸　　　（五）数均分子量（六）聚乙烯醇
（七）丙烯酰胺　（八）二甲基亚砜（九）脱保护
（十）元素　　　（十一）缩写　　（十二）链段
（十三）合成　　（十四）自由基　（十五）终止
（十六）引发剂　（十七）无规线团（十八）二甲基甲酰胺
（十九）提取　　（二十）杨氏模量

二、将下列中文词语译成英文（每小题 2 分，总分 10 分）

（一）monomer　　（二）homopolymer　（三）initiator
（四）conversion　（五）molecular weight

三、句子翻译（将英文句子翻译为中文，每小题 5 分，共 30 分）

（一）当我们提及支化聚合物，我们指的是来源于线性分子主链的附加的聚合链的存在。

（二）大多数单体的 k_p 值在 $10^2 \sim 10^4$ L/(mol·s)范围内。这是一个极大的数值，显著大于那些在平常化学反应中遇到的数值。

（三）这些化合物是由四氢呋喃的聚合反应制备的；它们也是线性的，而由环氧丁烷制备的化合物则以侧基为标记。

（四）离子聚合，类似于自由基聚合，也有链式反应的机理。然而，离子聚合的动力学与自由基聚合的动力学有很大的不同。

（五）到目前为止，包括原子转移自由基聚合和可逆加成-断裂链转移聚合等几种活性自由基聚合方法被相继报道。

（六）当我们讨论聚合物的分子量时，实际上论及的是它的平均分子量。

（要求内容正确，句子流畅通顺。词语错误或单词翻译错误每个扣 1 分，句子不通顺扣 1 分）

四、段落翻译（将英文翻译为中文，共 20 分）

聚合速率在实验上可以通过测定体系的几种性质的变化而确定，如密度、折射率、黏度或者吸光性能。密度的测量是这些技术中最准确最敏感的。对许多单体的聚合来说，密度增加了 20%～25%。在实际操作中，聚合体系的体积是通过在膨胀计中进行反应测定的。它是专门构造的带一个毛细导管的容器，由此可以对微小体积变化进行高精确度测量。通过膨胀计技术探测聚合过程中万分之几的变化是很常见的。

（要求内容正确，句子流畅通顺。词语错误或单词翻译错误每个扣 1 分，句子不通顺扣 1 分）

模拟测试题（四）答案及评分标准

五、写作（20 分）

一档（16~20 分）突出摘要的全部内容要点，包括实验目的、实验方法、表征方法简介和意义等内容。层次清楚，语言流畅，有句式的变化，有复杂结构，基本无语法错误或拼写错误。

二档（11~15 分）基本写出摘要的内容要点，层次清楚，语言流畅，存在少量的语法和拼写错误。

三档（6~10 分）包括了摘要一半以上的内容要点，语言不通顺，语法结构单一，只有少量句子完整，影响整体阅读。

四档（0~5 分）不扣题，词不达意，整体无法阅读。

其他要求及评分说明：字数不符合标准扣 2 分，书写潦草影响阅读扣 2 分，语法和拼写错误扣 0.5~5 分。

附 录

高分子领域常见外文 SCI 期刊信息汇总

期刊名称	数据库/出版商
Progress in Polymer Science	Elsevier
Polymer Reviews	Taylor and Francis
Journal of Membrane Science	Elsevier
Carbohydrate Polymers	Elsevier
Biomacromolecules	ACS
ACS Macro Letters	ACS
Macromolecules	ACS
International Journal of Biological Macromolecules	Elsevier
Polymer Chemistry	RSC
Macromolecular Rapid Communications	Wiley
Gels	MDPI
Cellulose	Springer
Polymer Degradation and Stability	Elsevier
Polymer	Elsevier
ACS Applied Polymer Materials	ACS
European Polymer Journal	Elsevier
Macromolecular Bioscience	Wiley
Macromolecular Materials and Engineering	Wiley
Polymers	MDPI
Membranes	MDPI
Polymer Testing	Elsevier
Chinese Journal of Polymer Science	Springer
Reactive & Functional Polymers	Elsevier
Soft Matter	Royal Society of Chemistry
Express Polymer Letters	BME-PT, Budapest University of Technology and Economics
Plasma Processes and Polymers	Wiley
Synthetic Metals	Elsevier
Journal of Polymers and the Environment	Springer
Polymer Journal	Springer
Polymers for Advanced Technologies	Wiley
Journal of Biomaterials Science—Polymer Edition	Taylor and Francis
Journal of Polymer Science	Wiley
Polymer International	Wiley
Journal of Applied Polymer Science	Wiley
Polymer Composites	Wiley
Journal of Polymer Science, Part B: Polymer Physics	Wiley
Macromolecular Chemistry and Physics	Wiley
Journal of Reinforced Plastics and Composites	Sage Publications

（续表）

期刊名称	数据库/出版商
Journal of Inorganic and Organometallic Polymers and Materials	Springer
Journal of Polymer Research	Springer
Journal of Cellular Plastics	Sage Publications
International Journal of Polymeric Materials and Polymeric Biomaterials	Taylor and Francis
Acta Polymerica Sinica	Science Press
Polymer Bulletin	Springer
Polymer-Plastics Technology and Materials	Taylor and Francis
Advanced in Polymer Technology	Wiley
Advances in Polymer Science	Springer
Polymer Engineering and Science	Wiley
International Journal of Polymer Science	Hindawi Limited
Macromolecular Research	Springer
Designed Monomers and Polymers	Taylor and Francis
International Journal of Polymer Analysis and Characterization	Taylor and Francis
Fibers and Polymers	Springer
European Physical Journal E	Springer
Colloid and Polymer Science	Springer
Journal of Bioactive and Compatible Polymers	Sage Publications
Iranian Polymer Journal	Springer
High Performance Polymers	Sage Publications
E-Polymers	De Gruyter
Rubber Chemistry and Technology	Rubber Division of the American Chemical Society
Macromolecular Theory and Simulations	Wiley
Macromolecular Reaction Engineering	Wiley
Green Materials	ICE Publishing
Journal of Vinyl & Additive Technology	Wiley
Plastics, Rubber and Composites	Taylor and Francis
Journal of Macromolecular Science Part A-Pure and Applied Chemistry	Taylor and Francis
Cellular Polymers	Sage Publications
Journal of Renewable Materials	Tech Science Press
Korea-Australia Rheology Journal	Springer
Journal of Elastomers and Plastics	Sage Publications
Polymers & Polymer Composites	Sage Publications
Journal of Macromolecular Science Part B-Physics	Taylor and Francis

（续表）

期刊名称	数据库/出版商
Journal of Polymer Engineering	De Gruyter
Polimeros-ciencia E Tecnologia	Associacao Brasileira de Polimeros
Polymer Science Series C	Springer
Polymer Science Series A	Springer
Mechanics of Composite Materials	Springer
Polymer Science Series B	Springer
International Polymer Processing	De Gruyter
Progress in Rubber Plastics and Recycling Technology	RAPRA Technology
Journal of Photopolymer Science and Technology	The Society of Photopolymer Science and Technology(SPST), Japan
Nihon Reoroji Gakkaishi	The Society of Rheology, Japan
Journal of Fiber Science and Technology	Society of Fiber Science and Technology
Polymer-KOREA	Polymer Society of Korea
Sen-i Gakkaishi	The Society of Fiber Science and Technology, Japan

生词表（Vocabulary）

A

abbreviate	v.	缩写，使省略
abbreviated	adj.	缩写的
abbreviation	n.	缩写，缩略词
ability	n.	能力
academic	adj.	学术的
accelerator	n.	促进剂
acetic acid		乙酸
acetone	n.	丙酮
acid	n.	酸
acidic	adj.	酸性的
acrylamide	n.	丙烯酰胺
acrylate	n.	丙烯酸酯
acrylic	adj.	丙烯酸的
acrylic acid		丙烯酸
acrylon	n.	腈纶
acrylonitrile	n.	丙烯腈
acrylonitrile-butadiene-styrene (ABS)		丙烯腈-丁二烯-苯乙烯共聚物
addition	n.	加入，加成
additive	n.	添加剂
adhesive	n.	黏合剂
alginate	n.	褐藻胶
alkali	n.	碱
alkaline	adj.	碱性的
allyl	n.	烯丙基
aluminum	n.	铝
amino	n.	氨基
amorphous	adj.	无定形的，非晶形的
amorphous region		非晶区
amylopectin	n.	支链淀粉
amylose	n.	直链淀粉
analogue	n.	类似物
anion	n.	阴离子
anionic	adj.	阴离子的
anhydrous	adj.	无水的
application	n.	应用
applicability	n.	适用性

生词表（Vocabulary）

architecture	n.	构造
atactic	adj.	无规立构的
atactic polymer		无规立构聚合物
atom	n.	原子
atomic force microscope		原子力显微镜
A_xB_y-type monomer		A_xB_y型单体
azobisisobutyronitrile	n.	偶氮二异丁腈

B

barrier	n.	壁垒
be characterized by		以……为特征
beam	n.	光线，横梁
benzene	n.	苯
benzoyl	n.	苯甲酰
benzoyl peroxide		过氧化苯甲酰
bifunctional	adj.	双官能团的
bifunctional monomer		双官能[基]单体
biocompatibility	n.	生物相容性
biodegradable	adj.	可生物降解的
biodegradable polymer		可生物降解的高分子
biopolymer	n.	生物高分子
blow	n.	强风
	v.	吹
blow molding		吹塑成型
blueprint	n.	蓝图
boundary	n.	边界
bond	n.	键
boron	n.	硼
branched	adj.	支化的
branched polymer		支化聚合物
bulk modulus		体积模量
bulk polymerization		本体聚合
butanol	n.	丁醇
butyl	n.	丁基
butyl chloride		丁基氯

C

calcium	n.	钙
calendering	n.	压延

calendering molding		压延成型
capability	n.	能力
carbohydrate polymer		碳水化合物聚合物
carbon	n.	碳
carbon (free) radical		碳自由基
carbon tetrachloride		四氯化碳
carbonyl	n.	羰基
carboxyl	n.	羧基
catalyst	n.	催化剂
catalyze	v.	催化
cationic natural polymer		阳离子天然大分子
cellulose	n.	纤维素
centigrade	n.	摄氏度（℃）
	adj.	摄氏度的
centimeter	n.	厘米
centrifugation	n.	离心分离
centrifuge	v.	使……受离心作用
	n.	离心机
certification	n.	证明
chain	n.	（分子）链
chain polymerization		链式聚合
chain reaction		连锁反应，链式反应
character	n.	特性
characteristic	n.	特征，特点
	adj.	独特的，典型的
characteristic equation		特性方程
characteristically	adv.	典型地
characterization	n.	表征
characterization factor		特性参数
characterize	v.	表征，描述
characterized	adj.	以……为特点的
chemical configuration		化学构型
chemical element		化学元素
chemical product		化工产品
chemical reaction		化学反应
chemical symbol		化学符号
chemical synthesis		化学合成
chiral	adj.	手性的
chirality	n.	手性
chitin	n.	甲壳素

生词表（Vocabulary）

chitosan	*n.*	壳聚糖
chlorine	*n.*	氯
chlorobenzene	*n.*	氯苯
chloroform	*n.*	氯仿
chromatographic	*adj.*	色谱法的，色析法的
chromatographically	*adv.*	色谱法地，色析法地
chromatographic column		色谱柱
chromatography	*n.*	色谱法
cis	*adj.*	顺式的
cis-1,4-polyisoprene		顺式聚异戊二烯
cis-form		顺式
coating	*n.*	涂层
coefficient	*n.*	系数
coil	*n.*	线圈
column	*n.*	柱
combination	*n.*	偶合
compatibility	*n.*	相容性
compound	*n.*	化合物
compression	*n.*	压缩
compression molding		模压成型
conduct	*v.*	实施，进行
configuration	*n.*	构型
conformation	*n.*	构象
conformational property		构象性质
conversion	*n.*	转化
conversion rate		转化率
copolymer	*n.*	共聚物
coupling reaction		偶联反应
covalent	*adj.*	共价的
covalently	*adv.*	共价地
crystal cell		晶胞
crystal morphology		晶体形貌
crystalline region		晶区
crystallinity	*n.*	结晶度
crystallization temperature		结晶温度
curve	*n.*	曲线
cyclic	*adj.*	环的
cyclohexane	*n.*	环己烷
cyclopentane	*n.*	环戊烷

D

data	n.	数据
data analysis		数据分析
data base		数据库
data processing		数据处理
D-configuration		D 构型
dead polymer		死聚合物
deformation	n.	形变
degas	v.	脱气
degassing	n.	脱气
degradable	adj.	可降解的
degradable plastic		可降解的塑料
degradable polymer		可降解的高分子
degradable property		可降解性能
degradation	n.	降解
degradation mechanism		降解机制
degradation product		降解产物
degradation rate		降解速率
degrade	v.	降解
degraded	adj.	已降解的
degree	n.	度
degree centigrade		摄氏度
degree of polymerization		聚合度
degree Celsius		摄氏度
dendrimer	n.	树枝状聚合物
dendrite	n.	树枝状晶体
dendritic polymer		树枝状聚合物
density functional theory (DFT)		密度泛函理论
derivative	n.	派生物
developing solvent		展开剂
diagram	n.	图表
dialysis	n.	透析
dialysis tube/bag		透析袋
diameter	n.	直径
diester	n.	二酯
differential	n.	微分
differential scanning calorimetry		差示扫描量热法
difunctional	adj.	双官能团的
dimension	n.	维度
dimer	n.	二聚体

生词表（Vocabulary）

dimethyl formamide		二甲基甲酰胺
dimethyl sulfoxide		二甲基亚砜
dipolymer	*n.*	二元共聚物
disproportionation	*n.*	歧化
dissolution	*n.*	溶解
dissolve	*v.*	溶解
distill	*v.*	蒸馏
dynamic light scattering		动态光散射
dynamic mechanical analysis		动力学分析

E

elastic	*adj.*	弹性的
elastic modulus		弹性模量
electrical property		电性能
electron	*n.*	电子
electronic	*n.*	电的，电子的
element	*n.*	元素
elute	*v.*	洗提
emulsion	*n.*	乳液
emulsion polymerization		乳液聚合
end-to-end distance		末端距
energy	*n.*	能源，能量
enzyme	*n.*	酶
equation	*n.*	等式，方程式
equipment	*n.*	装备，设备
establish	*v.*	建立
ethanol	*n.*	乙醇
ethyl	*n.*	乙基
ethyl acetate		乙酸乙酯
ethyl ether		乙醚
ethylene	*n.*	乙烯
ethylene-vinyl acetate copolymer		乙烯-乙酸乙烯酯共聚物
exchange	*n.*	交换
	v.	交换
exclusion	*n.*	排除
experimental data		实验数据
extend	*v.*	拉伸
extension	*n.*	拉伸
extract	*v.*	提取，萃取
extraction	*n.*	提取，萃取

extrusion	n.	挤出
extrusion molding		挤出成型

F

facility	n.	设施，设备
Fenton reagent		芬顿试剂
ferment	v.	发酵
fiber	n.	纤维
fibrous crystal		纤维晶
figure	n.	图
film	n.	膜
filter	v.	过滤
filtration	n.	过滤
flask	n.	烧瓶
flexible	adj.	灵活的，柔韧的
flexural modulus		弯曲模量
fluorescence	n.	荧光
fluorescence spectrophotometer		荧光分光光度计
fluorine	n.	氟
foaming	adj.	起泡的
foaming molding		发泡成型
folded chain model		折叠链模型
formula	n.	公式
formyl	n.	甲酰基
Fourier transform infrared spectrometer		傅里叶变换红外光谱仪
fraction	n.	部分
fracture	n.	断裂，截面
fracture surface morphology		断口形貌
fragmentation	n.	分裂
free radical		自由基
free radical polymerization		自由基聚合反应
freely jointed chain		自由连接链
freely rotating chain		自由旋转链
freeze	v.	冷冻
freeze drying		冷冻干燥
fringed-micelle model		缨状微束模型
function	n.	功能，函数
functional	adj.	功能的
functional group		官能团
functionality	n.	功能

生词表（Vocabulary）

functionally	adv.	功能地

G

gas	n.	气体
gaseous	adj.	气态的
gauche conformation		扭曲构象
gel	n.	凝胶
gel permeation chromatography		凝胶渗透色谱
gelatin	n.	明胶
general character		一般特征
glass-transition temperature		玻璃化转变温度
glassy	adj.	玻璃的
glassy state		玻璃态

H

heat	v.	加热
helical	n.	螺旋形的
helical conformation		螺旋构象
heparin	n.	肝素
heptane	n.	庚烷
hexane	n.	己烷
histogram	n.	直方图，柱状图
homolysis	n.	均裂
homopolymer	n.	均聚物
hydrocarbon	n.	碳氢化合物
hydrodynamic	adj.	流体的，流体力学的
hydrodynamic radius		流体动力学半径
hydrogel	n.	水凝胶
hydrogen	n.	氢
hydrolysis	n.	水解作用
hydrolytic	adj.	水解的
hydrolytically	adv.	按水解方法地
hydrothermal synthesis		水热合成
hydrothermal synthesis reactor		水热合成反应釜
hydroxy	adj.	羟基的
hydroxyl	n.	羟基

I

implement	*n.*	工具，器具
impurity	*n.*	杂质
in vacuum		真空中
inch	*n.*	英寸
incubate	*v.*	孵化，培育
inelastic	*adj.*	没有弹性的
information	*n.*	信息
infrared	*n.*	红外线
	adj.	红外的
inhibitor	*n.*	阻聚剂
inimer	*n.*	引发剂型单体
initiate	*v.*	引发
initiation	*n.*	（链）引发
initiator	*n.*	引发剂
injection	*n.*	注射
injection molding		注射成型
insoluble	*adj.*	不溶解的
instrument	*n.*	仪器，设备
integral	*n.*	积分
integral symbol		积分符号
intensely stir		剧烈搅拌
intensity	*n.*	强度
interface	*n.*	界面
interfacial	*adj.*	界面的
interfacial polymerization		界面聚合
intrinsic	*adj.*	内在的，固有的
intrinsic viscosity		特性黏度
ionic polymerization		离子聚合
iron	*n.*	铁
isocyanate	*n.*	异氰酸酯
isomerization	*n.*	异构化
isomery	*n.*	同分异构
isotactic	*adj.*	全同立构的
isotactic polymer		全同立构聚合物
isotaxy	*n.*	全同立构
italic	*n.*	斜体
ivory	*n.*	象牙

L

L-configuration		L 构型
lactic acid		乳酸
lamella	*n.*	片晶
laser	*n.*	激光
layer	*n.*	层
lignin	*n.*	木质素
linear polymer		线型聚合物
linkage	*n.*	链
liquid	*n.*	液体
	adj.	液态的
liquid crystal		液晶
liter	*n.*	升
lithium	*n.*	锂
living radical polymerization		活性自由基聚合
longitudinal	*adj.*	纵向的
loss modulus		损耗模量

M

macromolecular	*adj.*	大分子的
macromolecule	*n.*	大分子
main character		主要特征
manganese	*n.*	锰
manmade	*adj.*	人造的
	n.	人工制品
material characterization		材料表征
matrices	*n.*	基质（matrix 的复数形式）
matrix	*n.*	基质
mean square end-to-end distance		均方末端距
mean square radius of gyration		均方回转半径
mechanical property		力学性能
medium	*n.*	介质
melting temperature		熔融温度
membrane	*n.*	膜
metallic element		金属元素
methanol	*n.*	甲醇
methyl	*n.*	甲基
methyl acrylate		丙烯酸甲酯
methyl methacrylate		甲基丙烯酸甲酯

methacrylate	n.	甲基丙烯酸酯
methylene	n.	亚甲基
methylene chloride		二氯甲烷
micromorphological	adj.	微观形貌的
micromorphologically	adv.	微观形貌地
micromorphology	n.	微观形貌
microporosity	n.	孔隙率
microporous	adj.	多微孔的
milligram	n.	毫克
milliliter	n.	毫升
miscibility	n.	可混合性
mix	v.	混合
mixture	n.	混合物
model	n.	模型
modified natural polymer		改性天然高分子
moduli	n.	模（modulus 的复数形式）
modulus	n.	模量
molecular	adj.	分子的
molecular weight		分子量
mole per liter		摩尔/升
monoclinic syngony		单斜晶系
monocrystal	n.	单晶
monofunctional	adj.	单官能团的
monomer	n.	单体
monomeric	adj.	单体的
mole	n.	摩尔
morphological	adj.	形貌的
morphologically	adv.	形貌地
morphological character		形态特征
morphological characterization		形貌表征
morphology	n.	形貌，形态学
mould	n.	模具
multifunctional	adj.	多官能团的

N

nanosuspension	n.	纳米混悬液
naphthalene	n.	萘
natural	adj.	天然的
natural polymer		天然高分子
natural polymer derivative		天然高分子衍生物

English	POS	中文
natural polymer fibres		天然聚合纤维
natural polymer foam		天然高分子泡沫材料
nature	*n.*	本性
neutral	*adj.*	中性的
nitrogen	*n.*	氮
Nobel Prize		诺贝尔奖
non-natural polymer		非天然高分子
normalize	*v.*	标准化
nuclear magnetic resonance		核磁共振
number-average molecular weight		数均分子量
nylon	*n.*	尼龙

O

obtain	*v.*	获得，获取
o-dichlorobenzene		邻二氯苯
olefin	*n.*	烯烃
olefinic	*adj.*	不饱和的
oligomer	*n.*	寡聚体，低聚物
one-pot synthesis		一锅法合成
optical property		光学性质
organic solvent		有机溶剂
orientate	*v.*	取向
orientated	*adj.*	面向……的
orientation	*n.*	方向
original	*adj.*	原始的
original data		原始数据
oxidation reaction		氧化反应
oxidative degradation		氧化降解
oxygen	*n.*	氧
oxygen (free) radical		氧自由基

P

English	POS	中文
package	*n.*	包
packaging	*n.*	包装材料
paint	*n.*	漆
parameter	*n.*	参数
partial	*adj.*	部分的
particle	*n.*	颗粒
pattern	*n.*	模式

peak	n.	峰顶
peak molecular weight		尖峰分子量
peculiarity	n.	特点，特性
pellet	n.	小球
pentane	n.	戊烷
penetration	n.	穿透
percolation	n.	浸透
perform	v.	完成，执行，表现
performance	n.	性能，表现
periodic table of elements		元素周期表
permeation	n.	渗透
peroxide	n.	过氧化氢，过氧化物
perpendicular	n.	垂直
	adj.	垂直的
persulphate	n.	过硫酸盐
petroleum ether		石油醚
phenyl	n.	苯基
phosphorus	n.	磷
photoinitiation	n.	光引发
photoinitiator	n.	光引发剂
photosynthesis	n.	光合成
pioneering	adj.	首创的，先驱的
pitch	n.	沥青
plastic	n.	塑料
	adj.	塑料的
plot	n.	点状图
Poisson's ratio		泊松比
polar solvent		极性溶剂
poly(3-hydroxybutyrate)		聚-3-羟基丁酸（酯）
poly(butylene succinate)		聚丁二酸丁二醇酯
poly(lactic *co* glycolic acid)		聚乳酸羟基乙酸共聚物
polyacetaldehyde	n.	聚乙醛
polyacrylate	n.	聚丙烯酸酯
polyacrylate sodium		聚丙烯酸钠
polycaprolactone	n.	聚己内酯
polycrystal	n.	多晶
polydispersity	n.	多分散性
polyethylene	n.	聚乙烯
polyfunctional	adj.	多官能团的

生词表（Vocabulary）

polyglycolic acid		聚乙醇酸
polyhydroxyvalerate	n.	聚羟基戊酸酯
polyisoprene	n.	聚异戊二烯
polylactic acid		聚乳酸
polymer	n.	聚合物
polymer configuration		高分子构型
polymer product		高分子材料制品/聚合物制品
polymeric	adj.	聚合的
polymerization	n.	聚合
polymethyl methacrylate		聚甲基丙烯酸甲酯
polypropylene	n.	聚丙烯
polysaccharide	n.	多糖，多聚糖
polystyrene	n.	聚苯乙烯
polyvinyl alcohol		聚乙烯醇
polyvinyl chloride		聚氯乙烯
polyvinyl pyrrolidone		聚乙烯吡咯烷酮
pore	n.	毛孔
potassium	n.	钾
potential energy		势能，位能
precipitate	v.	沉淀
precipitation	n.	沉淀
preparation	n.	制备
processing	n.	加工处理，成型
product	n.	产物，产品
product quality		产品质量
profile	n.	轮廓图
propagate	v.	增长
propagation	n.	（链）增长
propanol	n.	丙醇
propene	n.	丙烯
property	n.	性能
propyl	n.	丙基
propylene	n.	丙烯
purification	n.	纯化，提纯
purify	v.	纯化，提纯
purity	n.	纯度
p-xylene	n.	对二甲苯
pyrene	n.	芘
pyridine	n.	吡啶

R

radical	n.	自由基
radical reaction		自由基反应
radical scavenger		自由基清除剂
radius	n.	半径
random	adj.	随机的
random coil		无规线团
rare earth element		稀土元素
rate	n.	比率，速率
raw data		原始数据
rayon	n.	人造丝
react	v.	反应
reactant	n.	反应物
reaction kinetics		反应动力学
reaction mechanism		反应机理
reaction rate		反应速率
reaction system		反应系统
reaction temperature		反应温度
reaction time		反应时间
reactive	adj.	反应的
reactive oxygen species		活性氧（种）
reactivity	n.	反应活性
reactivity ratio		竞聚率
reactor	n.	反应器
record	v.	记录，记载
	n.	记录，记载
recrystallization	n.	重结晶
redox	n.	氧化还原
reference	n.	参考
reflux	n.	回流
removal	n.	移除
repeating unit		重复单元
represent	v.	代表
residue	n.	残渣
resistance	n.	对抗
resolution	n.	分辨率
reversible	adj.	可逆的
revolutionary	adj.	革命性的，开创性的
rigid	adj.	刚性的
rigidity	n.	硬度

生词表（Vocabulary）

ring-opening		开环
rod	n.	棒
root-mean-square end-to-end distance		均方根末端距
rotation	n.	旋转
rough	adj.	粗糙的
rubber	n.	橡胶
rubbery state		橡胶态

S

sample	n.	样品
	v.	取样
scanning electron microscopy		扫描电子显微镜学
scheme	n.	方案
segmer	n.	链段
selectivity	n.	选择性
semicrystalline	adj.	半晶状的
separate	v.	（使）分离
separation	n.	分离
shake	v.	摇动
shaping	n.	成型
shear	v.	剪切
shear modulus		剪切模量
shish-kebab	n.	串晶
side reaction		副反应
silicon	n.	硅
smooth	adj.	平滑的
sodium	n.	钠
softener	n.	软化剂
solid	n.	固体
	adj.	固态的
solid-phase synthesis		固相合成
soluble	adj.	可溶的
solution	n.	溶液
solvent	n.	溶剂
spandex	n.	氨纶
spatial configuration		空间构型
species	n.	种类
spectra	n.	光谱（spectrum 的复数形式）
spectrum	n.	光谱
sphaerocrystal	n.	球晶

sphere	n.	球
square meter		平方米
starch	n.	淀粉
statistic	n.	统计数据
step-growth polymerization		逐步聚合
stereo-chemical configuration		立体分子构型
sticky	adj.	黏性的
stiff	adj.	僵硬的
stiffness	n.	硬度
stimulate	v.	刺激
stimulation	n.	刺激
stir	v.	搅拌
storage modulus		储能模量
strain	n.	张力，拉力
stress	v.	使……受压
	n.	压力
structure	n.	结构
structural	adj.	结构的
styrene	n.	苯乙烯
subscript	n.	下标
substance	n.	物质
sulfur	n.	硫
superscript	n.	上标
surface morphology		表面形貌
suspension	n.	悬浮液
swell	v.	溶胀
swelling	n.	溶胀
switchboard model		插线板模型
symbol	n.	符号，标志
symbol table		符号表
syndiotactic	adj.	间同立构的
syndiotactic polymer		间同立构聚合物
syndiotaxy	n.	间同立构
syngony	n.	晶系
synthesis	n.	合成
synthesize	v.	合成
synthesized	adj.	合成的
synthetic	adj.	合成的，人造的
	n.	合成物
synthetic fiber		人造纤维

生词表（Vocabulary）

synthetic material		合成材料
synthetic process		合成过程
synthetic reaction		合成反应
synthetic resin		合成树脂
synthetic rubber		合成橡胶

T

table	n.	表格
tacticity	n.	立构规整性
tank reactor		反应釜
temperature	n.	温度
tensile modulus		拉伸模量
terminate	v.	终止
termination	n.	（链）终止
terpolymer	n.	三元共聚物
tetrahydrofuran	n.	四氢呋喃
textile	n.	纺织物
thermal degradation		热降解
thermal gravimetric analyzer		热重分析仪
thermal property		热性能
thick	adj.	厚的
thickness	n.	厚度
thin layer chromatography		薄层色谱法
three-dimension (3D)		三维
toluene	n.	甲苯
trace element		微量元素
trans	adj.	反式的
trans-1,4-polyisoprene		反式聚异戊二烯
transverse	adj.	横向的
trans conformation		反式构象
trans-form		反式
transmission electron microscopy		透射电子显微术
trifluoroacetic acid		三氟乙酸
trimer	n.	三聚体
two-dimension (2D)		二维

U

ultrasonic	n.	超声波

	adj.	超声的
ultraviolet	*n.*	紫外线
	adj.	紫外的
ultraviolet and visible spectrophoto meter		紫外可见分光光度计
uniaxial	*adj.*	单轴的
uniform	*adj.*	均一的
unknown	*n.*	未知量
unparalleled	*adj.*	独特的，无与伦比的
unperturbed chain		无扰链
unsaturated	*adj.*	不饱和的

V

vacuum	*n.*	真空
value	*n.*	价值
variate	*n.*	变量
variation	*n.*	变化
vehicle	*n.*	工具
vinyl	*n.*	乙烯基
vinylacetate	*n.*	乙酸乙烯酯
vinyl-based monomer		乙烯基单体
vinylchloride	*n.*	氯乙烯
viscosity	*n.*	黏度
viscosity-average molecular weight		黏均分子量
viscous	*adj.*	黏的
viscous flow state		黏流态

W

weight	*n.*	重量
weight-average molecular weight		重均分子量

X

X-ray diffraction		X 射线衍射
X-ray photoelectron spectroscopy		X 射线光电子能谱（学）

Y

yield	*n.*	收率
Young modulus		杨氏模量

Z

zigzag	*n.*	锯齿形
	adj.	锯齿形的
zigzag conformation		锯齿构象

参考文献

[1] 魏无际, 俞强, 崔益华, 等. 高分子化学与物理基础. 2 版. 北京: 化学工业出版社, 2011.

[2] 曹同玉, 冯连芳, 张菊华, 等. 高分子材料工程专业英语. 2 版. 北京: 化学工业出版社, 2005.

[3] 周达飞, 唐颂超. 高分子材料成型加工. 北京: 中国轻工业出版社, 2006.

[4] 施良和, 胡汉杰. 高分子科学的今天和明天. 北京: 化学工业出版社, 1994.

[5] 董建华, 张希, 王利祥. 高分子科学学科前沿与展望. 北京: 科学出版社, 2011.

[6] Sarac A S. Redox polymerization. Progress in Polymer Science, 1999, 24: 1149-1204.

[7] Mülhaupt R. Hermann Staudinger and the Origin of Macromolecular Chemistry. Angew. Chem. Int. Ed., 2004, 43: 1054-1063.

[8] Hu C, Chen W, Yao H, et al. Facile Synthesis of Ladder-Type Polyacenes with Perylene-Fused-Pyrene Structures. Macromolecular Chemistry and Physics, 2018, 219: 1800201.

[9] Wang Y, Maeda R, Kali G, et al. Synthesis of Poly(Methyl Methacrylate)-Based Polyrotaxane via Reversible Addition-Fragmentation Chain Transfer Polymerization. ACS Macro Letters, 2020, 9: 1853-1857.

[10] Xue B, Cui W, Zhou S, et al. Facile Preparation of Highly Alkaline Stable Poly(arylene-imidazolium) Anion Exchange Membranes through an Ionized Monomer Strategy. Macromolecules, 2021, 54: 2202-2212.

[11] Liu F, Wu Y, Bai L, et al. Facile Preparation of Hyperbranched Glycopolymer via AB$_3$* Inimer Promoted by a Hydroxy/Cerium (Ⅳ) Redox Process. Polymer Chemistry, 2018, 9: 5024-5031.

[12] Shi J, Zhang H, Zhang B, et al. Halloysite nanotube-based nest-like composite microspheres with enhanced microwave absorption ability. Composites Science and Technology, 2022, 230: 109760.

[13] Ren X, Zhang H, Song M, et al. One-Step Route to Ladder-Type C-N Linked Conjugated Polymers. Macromolecular Chemistry and Physics, 2019, 9: 1900044.

[14] Zhang H, Cheng C, Song H, et al. A Facile One-Step Grafting of Polyphosphonium onto Halloysite Nanotubes Initiated by Ce(Ⅳ). Chemical Communications, 2019, 55: 1040-1043.

[15] Strain H H, Sherma J. Michael Tswett's contributions to sixty years of chromatography. J. Chem. Educ., 1967, 44: 235.

[16] Mülhaupt R. Catalytic Polymerization and Post Polymerization Catalysis Fifty Years After the Discovery of Ziegler's Catalysts. Macromolecular Chemistry & Physics, 2003, 204: 289-327.

[17] Lee K W, Chung J W, Kwak S, et al. Synthesis and characterization of bio-based alkyl terminal hyperbranched polyglycerols: a detailed study of their plasticization effect and migration resistance. Green Chemistry, 2016, 18: 999-1009.

[18] Hamada T, Yoshimura K, Takeuchi K, et al. Synthesis and Characterization of 4-Vinylimidazolium/Styrene-Cografted Anion-Conducting Electrolyte Membranes. Macromolecular Chemistry and Physics, 2021, 222: 2100028.

[19] Liu C, Sun M, Zhang B, et al. Synthesis and characterization of bisphthalonitrile-terminated polyimide precursors with unique advantages in processing and adhesive properties. Polymer, 2021, 212: 123290.

[20] Liu Y, Wei J, Luo Y, et al. The percolation behavior and positive vapor coefficient peculiarity (PVC) were investigated, and the effect of different ways of thermal treatments on the PVC peculiarity was also explored. Materials Research Innovations, 2006, 10: 52-57.

[21] Tan J, Chen W, Guo J, et al. The kind of polymers has shown the synthetic variety, the advanced capability and the wide applicability in contrast to the reported analogues. Chinese Chemical Letters, 2016, 27: 1405-1411.

[22] Takata T. The synthesis and dynamic nature of macromolecular systems controlled by rotaxane macromolecular switches are introduced to discuss the significance of rotaxane linking of polymer chains and its topological switching. ACS Central Science, 2020, 6: 129-143.

[23] Lu Z, Kong X, Zhang C, et al. Effect of colloidal polymers with different surface properties on the rheological

property of fresh cement pastes. Colloids and Surfaces A: Physicochemical and Engineering Aspects, 2017, 520: 154-165.

[24] Chen Y, Tian H, Yan D, et al. Conjugated Polymers Based on a S- and N-Containing Heteroarene: Synthesis, Characterization, and Semiconducting Properties. Macromolecules, 2011, 44: 5178-5185.

[25] Wang R, Liu W, Fang L, et al. Synthesis, characterization, and properties of novel phenylene-silazane-acetylene polymers. Polymer, 2010, 51: 5970-5976.

[26] Ballauff M. Nanoscopic polymer particles with a well-defined surface: Synthesis, characterization, and properties. Macromolecular Chemistry & Physics, 2003, 204: 220-234.

[27] Hung K, So F, Ken N, et al. Elastic modulus of ultrathin polymer films characterized by atomic force microscopy: The role of probe radius. Polymer, 2016, 87: 114-122.

[28] Zahedi F, Amraee I A. Carboxylated multiwalled carbon nanotubes effect on dynamic mechanical behavior of soft films composed of multilayer polymer structure. Polymer, 2018, 151:187-196.

[29] Zheng J, Ozisik R, Siegel R W, et al. Disruption of self-assembly and altered mechanical behavior in polyurethane/zinc oxide nanocomposites. Polymer, 2005, 46: 10873-10882.

[30] Brovko S, Palamarchuk A, Boitsova A, et al. Influence of the conformation of biopolyelectrolytes on the morphological structure of their interpolymer complexes. Macromolecular Research, 2015, 23: 1059-1067.

[31] Li Y, Xu T, Ouyang Z, et al. Micromorphology of macromolecular superabsorbent polymer and its fractal characteristics. Journal of Applied Polymer Science, 2009, 113: 3510-3519.

[32] Zhang H, Ren T, Ji Y, et al. Selective Modification of Halloysite Nanotubes with 1-Pyrenylboronic Acid: A Novel Fluorescence Probe with Highly Selective and Sensitive Response to Hyperoxide. ACS Applied Materials & Interfaces, 2015, 7: 23805-23811.

[33] Zhang H, Ren T, Yu M, et al. Synthesis and Characterization of Curcumin-Incorporated Glycopolymers with Enhanced Water Solubility and Reduced Cytotoxicity. Macromolecular Research, 2016, 24, 655-662.

[34] Wu Y, Yao K, Nie H, et al. Confirmation on the compatibility between cis-1,4-polyisoprene and trans-1,4-polyisoprene. Polymer, 2018, 153: 271-276.

[35] Wu Y, Hao X, Wu J, et al. Pure Blue-Light-Emitting Materials: Hyperbranched Ladder-Type poly(p-phenylene)s Containing Truxene Units. Macromolecules, 2010, 43: 731-738.

[36] Jiang Q, Zhang B, Yang J, et al. Investigation on structural changes of isotactic polypropylene mesophase in the heating process by using two-dimensional infrared correlation spectroscopy. Chinese Chemical Letters, 2015, 26: 197-199.

[37] Shi J, Zhang H, Wang Y, et al. Aminated Halloysite/Jute/Phenol-Formaldehyde Composites with Enhanced Mechanical Properties, Improved Lightweight Character, and Lowered Fire Hazard. ACS Applied Engineering Materials, 2023, 1: 1575-1582.

[38] Zhang B, Li S, Wang Y, et al. Halloysite nanotube-based self-healing fluorescence hydrogels in fabricating 3D cube containing UV-sensitive QR code information. Journal of Colloid and Interface Science, 2022, 617: 353-362.

[39] Lu Y, Zhao H, Huang X, et al. Exploring maleimide-anchored halloysites as nano-photoinitiators for surface-initiated photografting strategies. Chemical Communications, 2022, 58: 13636-13639.